Aging, Creativity, and Art
A Positive Perspective on
Late-Life Development

The Plenum Series in Adult Development and Aging

SERIES EDITOR:
Jack Demick, *Clark University, Worecester, Massachusetts*

A Continuation Order Plan is available for this series. A continuation order will bring
delivery of each new volume immediately upon publication. Volumes are billed only
upon actual shipment. For further information please contact the publisher.

Aging, Creativity, and Art
A Positive Perspective on Late-Life Development

Martin S. Lindauer
Professor Emeritus, State University of New York,
College at Brockport, Brockport, New York

Kluwer Academic / Plenum Publishers
New York ● Boston ● Dordrecht ● London ● Moscow

Library of Congress Catalong-in-Publication Data

Lindauer, Martin S.
  Aging, creativity, and art: a positive perspective on late-life development/
  Martin S. Lindauer.
    p. cm. – (The Plenum series in adult development and aging)
  Includes bibliographical references and index.
  ISBN 0-306-47756-4
    1. Creative ability in old age. 2. Creation (Literary, artistic, etc.) 3. Aged artists.
  4. Aged-Psychology. I. Title. II. Series.

  BF724.85.C73L56 2003
  155.67′1335–dc21

                                                                    2003051346

ISBN 0-306-47756-4

© 2003 Kluwer Academic/Plenum Publishers, New York
233 Spring Street, New York, New York 10013

http://www.wkap.nl

10 9 8 7 6 5 4 3 2 1

A CIP record for this book is availble from the Library of Congress

Permissions for books published in Europe: *permissions@wkap.nl*
Permissions for books published in the United States of America:
permissions@wkap.com

Printed in the United States of America

Preface

Share with me the thoughts that might be evoked while looking at a self-portrait of Rembrandt as an old man (see cover). You might ask yourself, for example, what was growing old like for him? More personal questions may be prompted: Is his lined face a reminder of my feelings about growing old? Look at his eyes. Do they show the suffering or strength of old age? Broader questions might follow: Can a portrait by a single aging artist, albeit a great one, tell me something about aging in general? Did other old artists depict aging more or less the same way? The questions expand: Are the elderly today viewed differently than they were in the past? Your focus returns to the Rembrandt. Is there a hint of losses or gains in his creativity? Did Rembrandt's late-life creativity radically change? You move your gaze away from the painting and wonder, What do aging artists and the art of old age tell us about aging? "A great deal" resoundingly answer art historians, literary critics, scholars in the humanities, and others who attribute revelatory powers to art and find in it an insightful understanding of the human condition.

....

Readers will not be surprised at the implied link between "Creativity" and "Art" in *Aging, Creativity, and Art*, since the terms, along with "Artist," are used interchangeably. Their juxtaposition with "Aging" is also understandable in light of the Rembrandt exercise above. But perhaps most surprising will be these terms' placement within the scientific frameworks of psychology and gerontology. Science and art? The humanities and statistics? Rare, indeed, are interdisciplinary exchanges between empirical research, discourses from the humanities, examples from the arts, and testimony from artists. The synthesis is made possible by combining the strengths of the scientific method, including its penchant for quantification, the virtues of the humanities, with their emphasis on the individual and the unique, and the novel advantage points of the arts, with their felicitous forms of expressive communication. While much has been written about aging, creativity, and art, none integrate the three as *Aging, Creativity, and Art* does.

The book is primarily about old artists, late art, and elderly participants in the arts. But it is applicable to professionals studying late-life development, non-professionals looking ahead to retirement, helping professions seeking practical ways of enhancing the elderly's physical and mental well-being, and anyone cheered by an optimistic account of aging. Several kinds of readers will therefore find the book useful. They include scholars curious about scientific excursions into their domain, scientists skeptical about the place of art in empirical inquiry, practicing artists interested in connections between their work, science, and public issues, practitioners in art therapy, recreation, and leisure for the elderly attentive to research related to their fields, and policy-makers and administrators of programs on creativity and art for the aged. The book, most generally, is directed to readers concerned about aging and late-life growth, and maintaining if not enhancing youthful abilities into the later years.

To meet these goals, the stage is set with a critical review of perspectives on aging and creativity, and the role of the arts in clarifying both (Part I), followed by a comprehensive discussion of the multiple paths taken by late-life creativity, especially as these are illustrated in the arts (Part II). The issues raised in these chapters are closely examined in the main body of the book (Parts III–V). The pertinent scientific literature and relevant discussions in the humanities are summarized. They provide a context for my studies on aging artists, their works, and self-reports, as well as aging non-artists' receptivity to paintings and the arts in general. Specifically, these chapters examine the creative output of renowned historical artists over their lifetimes, including the ages at which they produced masterpieces (Chapters 6–8); self-reports by elderly contemporary artists on their late-life creativity and the reasons for continued excellence (Chapters 9 and 10); a special kind of late art, the old-age style, and its bearing on creativity (Chapters 11–14); and the reactions of young and old viewers to different kinds of paintings (Chapter 15) and their activities, attitudes, and interests in the arts (Chapter 16). The conclusion (Chapter 17) returns to the issues raised at the beginning of the book—the relationship between aging, creativity, and art—in light of the arguments, examples, and studies presented in these pages, and ends with an optimistic outlook on old age: Creativity continues late into life and in some cases is reinvigorated; mental abilities of the sort that underlie creativity, such as problem-solving, are maintained until old age; the elderly are capable and productive; and art has the potential of improving physical and mental health. These themes counter a pessimistic view of growing old, promote a hopeful perspective on aging for an increasingly elderly population, and offer practical guidelines for the aged's greater involvement in the arts. *Aging, Creativity, and Art* is therefore about the strengths and opportunities of old age as revealed by the accomplishments of aging artists, the productions of late artistic works, and the activities of elderly arts audiences.

....

A personal note. I was professionally reared with the traditional view of creativity, namely, that it declines after age 30 or so. The best predictor of adult creativity, it was held, was youthful precocity. Because this rule was based on a good deal of scientific evidence (see Chapter 3), I believed it. When I passed 30, I was still hopeful, since my "occupational age" was shortened by a two-year stint in the Army, study abroad for a year, and part-time student status. The years passed, as did my fortieth birthday. I became a professor, and increasingly doubtful about becoming a creative "mover and a shaker," I reluctantly accepted my reduced aspirations, stoically resigned myself to being a competent worker in a corner of my profession, and hoped to achieve a modest level of distinction.

In my 50s, I read informal accounts of aged artists who produced masterpieces, continued to be productive, and developed new forms of creativity. I wondered if similar signs of continued vigor, renewed excellence, and resurgent achievement might also occur among old non-artists not known for any particular remarkable qualities in their youth. Something like late-life creativity, I reasoned, might be manifested by changing occupations, careers, and professions, or starting a new one; setting new goals or redirecting old ones; moving on to new ways of thinking and modes of imagination; posing new questions and pursuing unfamiliar lines of inquiry; searching within oneself for submerged talents and recharging hidden interests; and behaving in ways that were not just different but better than earlier efforts.

I asked myself if a scientific study of the neglected topic of late-life artistic creativity might cast new light on late-life creativity in general, on maintaining if not increasing cognitive and other abilities, and on the emergence of unexpected traits in old age, like wisdom. Noteworthy accomplishments, I suspected, may not be limited to youth, falter in early adulthood, and cease by middle age—or reside in geniuses only.

Compelling anecdotes, insightful observations, informed discussions, and astute conjectures are interesting, but gain in value, in my view, if translated into hypotheses and confirmed by the systematic and quantitative procedures of science. Hence, this book. As I write these words, approaching 70, I am much more optimistic about the creative potential of old age than I was at 30.

....

Over 700 references on the arts, aging, and creativity were examined and several hundred are cited in the References. In a few instances where original articles could not be found or were not available, I relied on abstracts and summaries by others, and occasionally a poorly reproduced photocopy. Every effort was made to access vast amounts of material that cut across several disciplines and touched on many areas. Limited space, time, energy, and purpose forced me to be selective. Consequently, I have probably omitted a few works that might cast some doubt on my evidence, throw my arguments slightly askew, and raise a question or two about my interpretations. Hopefully, there are no more than a handful of omissions and distortions, and these do not detract from my main thesis. Readers' help in correcting any mistakes are welcomed and appreciated.

Despite the wide net I have cast, I claim only modest expertise in one small corner of the many complex areas covered by *Aging, Creativity, and Art*. I have tried to be fairly complete, clear, and unbiased, often relying on the words and studies of others. I have drawn heavily upon my own research (some of which is unpublished), greatly expanding and extensively re-writing it for non-specialized audiences. As an experimental psychologist, conditioned to stay close to the facts, I have strayed occasionally, albeit cautiously, in order to achieve a broader perspective. My speculations, though, are grounded in facts, and when these were not available, I have suggested research to support them.

I have provided the details about my research, including statistics, tables, and figures, so that their samples, methods, procedures, and materials are public and thereby open to alternative (and opposing) views that may counter my interpretations and fuel independent judgments. Overly technical matters, though, are in the footnotes to each chapter. By providing the specifics of research I also intended to illustrate a more general point: scientific methods, including quantification, can be applied to materials as subtle and subjective as art, the discussions of art historians, and the testimonies of artists without violating their spirit or distorting their intent.

Works are cited in the body of the text according to the American Psychological Association (APA) style, which cites only the author and the date of publication (e.g., Lindauer, 1990); the full citation is found in the References section at the end of the book. More specialized references to the literature on aging, art, and creativity are in the bibliographies of cited works. The note in Chapter 1 lists a fairly complete set of general and mostly recent readings in creativity. A broad reading list on aging, art, gerontology, and psychology will readily be found in college libraries. The quotations that head the Parts and Chapters are from standard references available in public libraries (e.g., Sampson & Sampson, 1985).

I come to this book as a teacher with over 25 years of experience, a state-wide teaching award, a Fellow of the Teaching Division of the American Psychological Association, a former President of the Psychology and Arts Division 10 (now The Society of the Psychology of Aesthetics, Creativity, and the Arts), and author of over 100 research articles, chapters, encyclopedia entries, essays, and papers on the arts, creativity, perception, and cognition, including *The Psychological Study of Literature* (Lindauer, 1975).

Much of the readings and research on which this book is based was initiated when I was a Senior Research Fellow, awarded by the National Institute on Aging, when I was on leave from the State University of New York, College at Brockport, where I am a Professor Emeritus, and was spent at the Institute on Aging at the University of Michigan. The full-time support provided by the Fellowship gave me a marvelous and once-in-a-lifetime opportunity to work intensively in the areas of aging, creativity, and art. The Institute on Aging, under the directorship of Dr. Richard C. Adelman, in addition to providing me with space, secretarial, and other resources, also offered numerous speakers, colloquia, and exchanges with colleagues that considerably enriched my knowledge of gerontology. The Psychology Department at Michigan, and its Chairman, Dr. Albert C. Cain, were also helpful in many ways, including allowing me to teach a course related to the work reported here and in providing administrative support for grant applications and research projects.

Throughout my two years at Michigan, Dr. Marion Perlmutter served both as professional mentor and personal guide through a new field of study for me, offering encouragement, direction, and gentle criticism. Useful, too, were discussions with Professor Perlmutter's graduate and post-graduate students in her seminars, including Karen Fingerman, A. J. Faulkner, and David Crystal, among others, several of whom collected and analyzed much of the data reported in Chapter 16; special thanks to Debra Vella, another member of the group, for the data and analysis she provided for the research reported in that chapter. The studies reported in Chapter 16 were a small segment of a much larger project supported, among others, by AARP (formerly American Association of Retired Persons). I am also especially appreciative of the friendship and help of Dr. Carolyn Adams-Price, also at the Institute during my tenure. My sincere thanks, as well, to Dr. Lucinda Orwoll, with whom I worked closely on the questionnaire study of artists, reported in several chapters, and to Dr. Carolyn Phinney, who were also at the Institute, as was Dr. Andrew Achenbaum, a bountiful spirit to whom I looked for assistance, laced with good humor and intelligence. Dr. Mark Runco, Editor of the *Creativity Research Journal* in which I published several of the studies reported here, was a source of technical advice and useful criticism, as was Dr. Colin Martindale, Editor of *Empirical Studies of the Arts*, in which some of the material reported here was first published.

The data for Chapter 15 were collected at Bowling Green State University; the cooperation of Drs. Robert Connors and Joseph Cranny of the Psychology Department are greatly appreciated. Thanks also to the art historians who participated in several studies (in Part IV) and for their tolerance of what must have seemed to them a crass example of scientific number-crunching.

Thanks, too, to my teachers at the University of Illinois and the New School for Social Research (now The New School University), Professors Asch, Greenbaum, Heimer, Henle, Rock, Wallach, and others who introduced me to Gestalt psychology. There are also many fellow members of Division 10 of the American Psychological Association to whom I have looked in admiration, with respect, and for support, including Frank Barron, Stephanie Dudek, Joseph Juhasz, Howard Gruber, Marge Franklin, Ravenna Helson, Nat Kogan, Pavel Machotka, Dean Simonton, Tobi Zausner, and others too numerous to list. Finally, for her sensitivity to the moods of my old age, her patience with my impatience, and her willingness to listen, give encouraging feedback, and offer friendly questioning, I am forever grateful to Bonnie.

I wish to thank Lawrence Erlbaum Associates, Mahwah, NJ 07430, for permission to adapt the tables in Lindauer (1993). The span of creativity. *Creativity Research Journal, 6*, 329–343, and for excerpts from Root-Bernstein, Bernstein, & Garnier (1993). Identification of scientists making long-term high-impact contributions, with notes on their method of working. *Creativity Research Journal, 6,* 329–343; and to Alinari/Art Resource, NY, and National Gallery, London. Great Britain, for permission to reproduce the Rembrandt *Self-Portrait as an Old Man,* 1669. For permission to quote materials I wish to thank the following: Academic Press, San Diego, CA, for excerpts from Root-Bernstein (1999). Productivity and age. In Runco & Spitzker (Eds.), *Encyclopedia of creativity* (vol. 2, pp. 457–463); Allen Memorial Art Museum, Oberlin College, Oberlin, OH, for excepts from Boggs (1977–78). Edgar Degas in old age. In Hood (Ed.), *Artists and old age: A symposium.* Allen Memorial Art Museum Bulletin Oberlin College, *35,* 57–67; American Psychological Association, for excerpts from Simonton (1989), The swan-song phenomenon: Last-work effects for 172 classical composers. *Psychology and Aging, 4,* 42–47; American Society on Aging, San Francisco, CA, for excerpts from Simonton (1991) Creative productivity through the adult years. In Kastenbaum (Ed.), Creativity in later life. *Generations: Journal of the American Society on Aging, 15,* 13–16; American Association for the Advancement of Science, for excerpts from Rowe and Kahn (1987). Human Aging. *Science, 237,* 143–149; Archon Books/The Shoe String Press, North Haven, CT, for excerpts from Rothenberg & Greenberg (1974). *The index of scientific writings on creativity;* Wayne Booth. for excerpts from Booth (1992). *The art of growing older: Writers on living and aging;* the Board of Trustees of the Leland Stanford Jr. University, with the permission of the Stanford University Press, www.sup.org, for excerpts from Holahan, et al. (1995). *The gifted group in later maturity,* copyright © 1995, Stanford, CA; Gerontological Society of America, Washington, DC, for excerpts from Rowe & Kahn (1997). Successful aging. *The Gerontologist, 37,* 433–440, and for excerpts from Lehman (1956). Reply to Dennis. *Journal of Gerontology, 11,* 333–337; Greenwood, Westport, CT, for excerpts from Achenbaum (1989), Foreword Literatures value in gerontological research. In Bagnell & Sapir (Eds.) *Perception of aging in literature: A cross-cultural perspective;* Guilford Publishers, NY, NY, for excerpts from Pruyser (1987). Creativity in aging persons. *Bulletin of the Menninger Clinic, 51,* 425–435; Haworth Press, Binghampton, NY, for excerpts from Birren (1990). Creativity, productivity, and potentials of the senior scholar. Exploring the feasibility of an institute of senior scholars [Special Issue]. *Gerontology and Geriatrics Education, 11,* 27–44; and for excerpts from Scotch (1989). Foreword. In McMurray, Creative arts with older people. *Activities, Adaptation and Aging, 14*(1/2), 1–138; Heldref Publications, the Helen Dwight Reid Educational Foundation, 1319 18th Street, NW, Washington, DC 20036-1802, for extracts from Lehman (1966). The most creative years of engineers and other technologists. *Journal of Genetic Psychology, 108,* 263–270; *Journal of General Psychology,* Washington, DC, for excerpts from Bink & Marsh (2000). Cognitive regularities in creative activity. *Review of General Psychology, 4,* 59–78; National Council on the Aging, National Center on the Arts and Aging, Washington, DC, for excerpts from Sunderland, Traylor, & Peter Smith Associates (Eds.) (1976/1977). *Arts and the aging. An agenda for action: A national conference,* and for excerpts from Moody (1982), *Aging and cultural policy;* National Gallery of Art, Washington, DC, for excerpts from Feldman (1992). *I am still learning: Late works by masters;* Princeton University Press, Princeton, NJ, for excerpts from Lehamn, *Age and Achievement* (1953); Springer, NY, NY, for excerpts from Birren (1985). Age, competence, creativity, and wisdom. In Butler, & Gleason (Eds.), *Productive aging: Enhancing vitality in later life* (pp. 29–34); Tayler & Francis, Philadelphia, PA, for excerpts from Sunderland (1982). The arts and the aging advocacy movement in the United States: A historical perspective. *Educational Geronotology, 8,* 915–205; *The New York Times,* NY, NY, for excerpts from Grumbach (2000). What old age is really like. *The New York Times, 148,* 15; the Newhouse News Service, for excerpts from Foote (1994). The art of aging gracefully. *The Oregonian,* August 14, D1, D4; and the Whitney Museum of American Art (in conjunction with the exhibition Enduring Creativity, Fairfield County, April 15–June 15, 1988), for excerpts from Feinstein (1988). *Enduring Creativity,* © 1988 Whitney Museum of American Art.

Contents

Part IV: The Old-Age Style

PART I

The Case is Made

Late-Life Creativity and Old Age Art

Grow old with me!
The best is yet to be.
The last of life, for which the first was made;
Our times are in his hand
Who saith, "A whole I planned,
Youth shows but half; trust God; see all,
nor be afraid!!"
Robert Browning, *Rabbi Ben Ezra*

CHAPTER 1

Late-Life Creativity

With the ancients is wisdom;
and in length of days understanding.
Old Testament, *Job*

The older segment of the population is growing faster than any other age group. In the United States, those over 65 are projected to account for 20% of the nation, reflecting a worldwide pattern (McDonald, 1997; Perlmutter, 1990). Average life expectancy has not only increased but has led to two new categories of old age, the very-old and the old-old. There are more than 664 million people over 80 in the world, 1 for every 100 on earth, a number predicted to grow by another 370 million in 2050. The 135,000 centenarians in the United States in 1998, say forecasters, will reach 2.2 million in the next 50 years. The upsurge of the aged is reflected in more than 1,000 college programs in gerontology, with several hundred at the graduate level, a 36-fold increase in funding from the National Institute on Aging in the last 25 years, to $685.6 million in a recent accounting, and more than a 10-times growth in the number of grants in the same period, valued at $722 million.

These changing demographics are bound to affect attitudes toward the old, including the elderly's views of themselves, and alter the meaning of growing old from what it was in the past. More people live longer, to over seventy, and many continue to work or go to school. The nature of retirement differs; senior citizen groups, many with militant agendas, are well organized, and laws are legislated against age discrimination (Chudacoff, 1992; Fischer 1977; Murray, Sleek, & DeAngeles, 1997; Qualls & Abeles, 2000). In response to the graying of the population, a large number of reports, essays, articles, and books on growing old have been written by psychologists, gerontologists, social scientists, historians, health specialists, and journalists (Cole, 1992a, 1992b; Cole & Winkler, 1994; Craik & Salthouse, 1992), contributing to the new fields of "humanistic gerontology" and "geropsychology" (Birren & Schroots, 2000). The government's interest in the aged was formally initiated in 1950 by the National Council on the Aging, followed by the National Center on the Arts and Aging in 1973 "to serve as an advocate and agent for linking the arts and the aging together...to ensure that older persons have an equal opportunity with other age groups to participate in and have access to the arts" (Sunderland, Traylor, & Peter Smith Associates, 1976/1977, p. 1).

Supplementing professional and governmental efforts are popular works on finances, retirement, housing, and recreation led by Simone de Beavoir's influential 1950 book, *The Coming of Age* (titled *La Vieillesse* in French, or *Old Age*), which drew upon literature, the arts, history, psychology and biology in a personal and readable way (Woodward, 1994), although it presented a view of aging that Achenbaum (1989) called grim (see pp. viii–xxv).

The many connotations of aging were formalized in McLerrran and McKee's (1991) *Cultural Dictionary*, which annotates 350 references to the elderly in the arts, literature, and culture. The number and diversity of citations, the authors contended, indicated that old age is an "archetype" that "resonates with richly ambiguous meanings" (p. ix). The multiple and changing historical, philosophical, humanistic, methodological, disciplinary, and substantive areas of aging are based on continuing series of handbooks (Birren & Schaie, 1977, 1985, 1990; see also Maddox, 1987).

OLD AGE: A TIME OF WOE OR WORTH?

Two opposing perspectives broadly characterize approaches to aging. First and foremost is an emphasis on the anxieties of old age: having enough money for food, housing, and medical care; coping with increasing illnesses, physical deterioration, and sensory debilities; facing the deaths of spouse, relatives, and friends; and coming to terms with one's mortality. To these worries add fears of mental decline and stagnation, losses in intellect and intelligence, impaired ability in learning and remembering, and the onset of senility or dementia, including Alzheimer's disease. These foreboding preoccupations, according to Cummings (1979), lead to the elderly's "disengagement" from life, a self-protective distancing from feelings, other people, and activities that result in personal and social isolation. A brief poem by W. H. Auden (1976), entitled *Old People's Home* (p. 645), captures pessimistic sentiments about aging.

According to the negative view of old age, losses take precedence over gains, bad things are more likely to happen than good ones, and despair triumphs over hope. A litany of woes pervade societal attitudes toward the old and the elderly's attitudes about themselves, and form the basis for stereotypes about aging and the prejudice of ageism.

Worse, this depressing profile is indirectly supported by research and theories of human development that pay less attention to the old than the young (e.g., childrearing, schooling, growing up, and adolescence), rarely looks past middle-age (Arlin, 1975), and fails to challenge the presumed disadvantages of aging, which are taken for granted, ignores alternatives, and slights benign possibilities. Implicit is the belief that whatever happens late in life has already been laid down in childhood and simply plays out with age. The "action" seemingly lies in the younger years, while nothing really new happens in the later decades when there are no fundamental changes comparable to youth.

The low opinion of old age in psychology may have begun with the pioneering investigations by Hall (1922, pp. 319–365), one of the founders of developmental psychology. In one of the first studies of old age, he sent a questionnaire to "a few score of eminent and distinguished old people, acquaintances and strangers." The questions anticipated current topics in old age as well as some not yet pursued but meriting inquiry. Some examples follow: The age at which respondents felt old; to what they ascribed their long life; how they kept healthy; regrets over things done or not; temptations felt; changes in interests and pleasures; preferences for the company of younger or same-aged people; and attitudes toward death and religion. Halls' impressionistic interpretations of the replies foreshadowed current negative positions, although he admitted they might have reflected the immanence of his own death.

A more recent and official statement by the American Psychological Association (Guidelines, 1998) on the relationship between aging and mental life also arrived at a muted but discouraging conclusion.

> Although some healthy aging persons maintain very high cognitive performance levels throughout life, most older people will experience a decline in certain cognitive abilities. This decline is usually not pathological.... For some older persons, however, declines go beyond what might be considered normal and are relentlessly progressive, robbing them of their memories, intellect, and eventually their abilities to recognize spouses or children.... (p. 1298)

To give this assertion some context, psychology generally takes a pessimistic orientation to other topics as well (Lindauer, 1968), although this stance has recently been questioned and revised under the rubric of "positive psychology" (Seligman & Csikszentmihalyi, 2000).

Critics of aging's disparagement question the cognitive superiority of youth over age in all cases (Arenberg, 1978; Giambra, 1987; Rodeheaver, 1987; Salthouse, 1987) and the reliability of age differences

(Graf, 1990; Roediger, 1990; Salthouse, 1989). Others look at (and for) the brighter side of growing old. The accumulated years of the elderly are not a burden but a source of growth and an opportunity to learn more about oneself, others, and the world. Birren (1985) pointed out that "assumptions about the marginality of older people [are] unsound [and] for people who have reasonably good health, there appears to be no negative relationship between age and productivity [especially when] productivity depends upon abstract skills rather than physical strength and endurance. Curiosity and intellectual vigor...frequently exist within a fragile body" (pp. 29–31). Taking an affirmative position, too, Butler and Gleason (1985) declared that with age there is a "growth in capacities for strategy, sagacity, prudence, wisdom, seasoning and experience" (pp. 7–8). Moody (1982a, p. 2) also objected to a narrow assessment of aging. "[A] preoccupation with the problems of late life can blind us to its triumph and its unexplored possibilities." Moody recognized aging's sickness, loneliness, and poverty, but also saw its opportunities. "We ignore latent strengths [of old people] and respond only to [their] dependency or failure."

A positive view of aging points out that training and practice can offset some mental losses. Thus, memory is better among aging university professors than those with less challenging positions (Shimamura et al., 1995); and old architects had fewer decrements in cognition than old nonarchitects (Salthouse, 1990) and their personalities did not change from midlife to old age (Dudek & Hall, 1991). For older working people, improved conditions, such as better lighting, larger type, and wider displays, helped them "stay productive [which indicates that] age-related declines in cognition [need] not lead to decays in job performance" (Azar, 1998, p. 23). Exercise, too, overcomes aging's losses (Ager, White, Mayberry, Crist, & Conrad, 1981–82).

Generalizations about decrements in old age therefore have to be qualified. They depend on the type of activity, the way it is investigated, such as the kinds of tasks used to gauge changes, whether individual differences are reported, and if the samples studied included the old as well as the very old, the healthy or the institutionalized. Birren (1985, p. 31) made a useful distinction between biological aging, or "senescing," which refers to losses and weaknesses; social aging, or "eldering," which emphasizes the roles assigned to the aged as a function of culture and society; and psychological aging, or "geronting," which denotes attitudes towards old age that influence perceptions, choices, decisions, and one's place in society. Senescing and its negative aspects are overemphasized, Birren contended, and eldering and geronting, which have either negative or positive outcomes, are neglected.

Many of the perils of aging—to be old is to be sick, frail, and nonproductive—have been reexamined, reevaluated, and in many cases replaced by more positive outcomes in a series of long-term and comprehensive studies by Rowe and Kahn (1987; all quotes are from p. 433; see also 1997, 1998). Their reappraisal, begun in the late 1980s, was based on over 1,000 "high functioning" older people, that is, individuals who were engaged with life, not ill, and functioning at good physical and cognitive levels. More than 100 studies of these healthy seniors revealed that losses reported in the past have either been exaggerated or can be attributed to shifts in "life style, habits, diet, and an array of psychosocial forces extrinsic to the aging process," as well as to education, economics, and historical factors independent of age. Rowe and Kahn concluded that "much of the cognitive loss in late middle life…may therefore be preventable [or] even be reversible." They also pointed out that "differences between age groups [ignore] the substantial heterogeneity within age groups," which means that some old people do better than their peers and even better than some younger individuals.

Proponents of the worth of old age, in opposition to those who see it as a time of woe, believe some abilities do not fade, and they reinterpret losses as independent of age, arguing that declines can be offset through practice, exercise, training, and improved working conditions. They also hold that compensations draw on remaining strengths, and new skills can be developed. As Butler (1974, p. 529) optimistically put it, "one can find use for what…has been obtained in a lifetime of learning and adapting." The positive outcomes of old age are labeled "successful (or optimum) aging" (Coleman, 1992; Fisher & Specht, 1999; Kastenbaum, 1994a; Neugarten et al., 1980; Schulz & Heckhausen, 1996; Wong, 1989).

The notion of successful aging expands traditional ways of thinking about aging as either normal (or usual) or abnormal (or pathological).

> Gerontology is broadening its perspective from a preoccupation with disease and disability to a more robust view [that is more than] the avoidance of disease and disability [but] the maintenance of high physical and cognitive function, and sustained engagement in social and productive activities. (Rowe & Kahn, 1987, p. 433)

Successful aging therefore debunks "the myths associated with old age" (Wetle, 1991, p. 3). Losses are not synonymous with growing old, the later decades of life are not necessarily impoverished, and there are viable alternatives to the inevitability of decline.

Successful aging, though, is not without its critics (Ryff, 1982). Success has many and vague meanings; it defines any age, not just old age; and it can refer to feelings and attitudes as well as behavior. How to achieve successful aging is problematic. What kinds of interventions and public policies might promote successful aging? Attempts to age successfully may turn out to be unsuccessful, as when older people take a lifestyle more suitable to youth. Not usually considered part of successful aging are creative activities and engagement in the arts, although Simonton (in Seligman & Csikszentmihalyi, 2000) hints at a link when he argues that creativity is "a special and significant form of optimal human functioning" (p. 151).

Criticisms of successful aging, however, do not detract from its positive contribution to an understanding of aging. Additional confirmation lies in longitudinal research, which tracks individuals from childhood to old age. The prototypical study is Terman's, who with colleague and successors traced the abilities and accomplishments of children with high IQ scores from the early part of the 20th century to their 70s; they are still being investigated (Cravens, 1992; Holahan, Sears, & Cronbach, 1995).

The intellectually gifted, when they became old, maintained considerable control over their lives, retained their sense of selfhood, interpreted life experiences meaningfully, improved their social relationships, achieved wisdom, and continued to be creative. On the bases of follow-up studies of the Terman group, as well as the longitudinal research of others, Holahan et al. concluded that old people's talents vary enormously, that being old is not the same for everyone, that some aspects of physical aging are preventable through exercise and nutrition, and that favorable environments can mitigate decline or minimize its impact.

A similarly sanguine account of personality development in old age characterizes longitudinal studies of Erikson, Erikson, and Kivnik (1986). The research was based on the recorded reflections of 29 octogenarians for whom life histories covering half a century were available. Following Erikson's conception of the stages of life, every decade poses tensions, challenges, and conflicts. These may be successfully overcome or not. In the case of the octogenarians, evidence was found for new attitudes, strengths, and virtues, the resolution of life-long crises, a freer expression of love and caring, and wisdom. Erikson and his colleagues also found stagnation, resignation, isolation, and apathy in some, but most showed integrity, intimacy, and hope. As mental and other functions began to decline as life neared its end, the 80-year olds looked back on the past and ahead to the future, and struggled to maintain their sense of selfhood by remaining "vitally involved." They did this by resolving issues of health, retirement, finances, family, marriage, friendship, work, and religion, as well as staying fully engaged in new activities, some of which included art and other creative modes of expression.

Not everyone agrees with Erikson's hopeful views of aging. "Realistically speaking," Clayton (1975) rhetorically asked, "are there any wise elderly men?" to which she bleakly answers, "Very few." (p. 120). She also found little empirical evidence in support of Erikson's stages of aging and judges the data less convincing than the theory. Echoing these sentiments in a more popular context, the writer and literary critic Doris Grumbach (1998, pp. 15, A30), in an especially dour op-ed piece in the *New York Times* following Senator John Glenn's second expedition into space at age 77, and under the heading "What old age is really like," objected to the idea of old age being the best of times. As part of her argument, she disagreed with the poet Robert Browning who had Rabbi Ben Ezra say in a poem, "Grow old along with me. The best is yet to be." Instead, she agreed with Emerson that being old is "an incurable disease" and sadly noted that only "a happy few escape this general condition [of loss and only] a tiny number...write or paint or are creative in some way. A few elderly people (but surely not many) do their best work in those years. A small minority maintain their vigor of mind without the assault of affliction or deterioration." She approvingly quotes the mother of Oscar Wilde: "Life is agony and hope, illusion and despair, all commingled, but despair outlasts all."

Two sharply opposed positions on aging emerge from these discussions: it is either a time of woe or worth. A more subtle variant: The old may be as good as they once were, but compared to youthful individuals with vastly different kinds of life experiences, their thinking and actions are different—not worse. What seems "bad" reflects, according to youngsters, old fashioned and outdated ways of expressing themselves and behaving. But from the oldsters' perspective, who either do not know or care about the latest fads and fashions, it is the youthful point of view that attributes to them reduced abilities and serious inadequacies. Declines, losses, and deteriorations among the elderly are therefore the result of comparisons with the young rather than with the seniors' youthful selves. The elderly have a good old age—according to them, not younger observers. Unfortunately, older people often use the standards of younger generations to evaluate their thinking and behavior rather than their own criteria.

Another option takes a mixed view of old age: It is both good and bad. Yes, there are losses with aging, but also stability and gains;

declines exist, but vary in degree; there are important exceptions; and serious losses do not occur until very late in life (Bergmann, 1972; Denny, 1982; Horn, 1982; Montagu, 1989; Schaie & Stroller, 1968). A balanced compromise between the positive and negative aspects of aging does not deny the reality or even the predominance of the latter. But by recognizing two (or more) possibilities, old age need not be solely or mainly a time of woe.

The more pessimistic alternative, though, has taken front stage, making it difficult to alter the overriding perception of aging as a time of doom and gloom. Consider the status of wisdom, arguably the most positive aspect of aging. Nevertheless, research "has tended to neglect those qualities [like wisdom] that might be unique to, and representative of, late adulthood" (Baltes, Smith, Staudinger, & Sowarka, 1990, pp. 129–130; see also Baltes & Staudinger, 1995, 2000; Clayton, 1987). Happiness, too, is another unexpected and unexplored gain of old age (Rabasca, 1999, based on a forthcoming study by Mroczek and Kolarz). "[O]lder adults regulate their emotions more effectively than younger or middle-aged adults [knowing] through their years of experience, what kind of external events increase or decrease their positive and negative emotions [thereby allowing them to direct] their lives toward maximizing positive affect and minimizing negative affect" (p. 11). Neglected, too, is another positive outcome of aging: late-life creativity.

LATE-LIFE CREATIVITY

Views of aging range from bleak pessimism to tempered optimism. Controversy, too, surrounds the topic of creativity. Are creative people mad? Is everyone creative? Is creativity innate or learned? How are creativity and intelligence related? Is creativity a personal or cognitive trait? A process? A behavior? Is creativity the same as new, original, unique, and valuable? How much of a contribution does society and culture make to an individual's creativity? Particularly vexing is whether there is more than one kind of creativity (Baer, 1999a; Gardner, 1983; Solomon, Powell, & Gardner, 1999), differing for scientists, various kinds of artists, and others. Creativity is also partitioned into other forms: spontaneous, expressive, productive, inventive, innovative, and original (Taylor, 1959). An attempt to organize the many aspects of creativity places them under one of four "P's": Process, person, product, press (environment; Richards, 1999b).

These basic definitional questions about creativity are not close to resolution. No "consensual understanding [exists of this] highly complex and diffuse construct," complained Mumford and Gustatson (1988, p. 27). Creative "behaviors have eluded a satisfying concrete definition [and] little agreement exists among academic disciplines as to the central attributes of creativity [because] empirical research on creativity has not converged on rules or definitions that identify performance as creative or not," concluded Bink & Marsh (2000, p. 59; see also Sternberg, 1985). Symptomatically, the definition chapter in a recent handbook on creativity spanned 13 pages while other topics took up two or three (Cropley, 1999; see also Ruth & Westerlund, 1979).[1] Even artists, who have a close and daily engagement with creativity, do not find it "an issue of much interest" (Rodeheaver, Emmons, & Powers, 1998, p. 203; cf. Tomas, 1964). Thus, the painter Barnett Newman humorously maintained that "aesthetics [creativity] is for artists like ornithology is for the birds." In their discussion of aging, Levy and Langer (1999) defined creativity as "the ability to transcend traditional ways of thinking by generating ideas, methods, and forms that are meaningful and useful to others" (p. 45). Note, though, that the key terms require further explication: Who decides by what criteria which ideas are "meaningful" and "useful"? And who are these "others" who make judgments about creativity?

---

[1] Extended discussions of the content, issues, and controversies of various approaches to and theories of creativity are found in numerous books. For a brief introduction, see Getzels (1990). For reviews of the field, see Kogan (1987) and Simonton (1999a). A collection of articles edited by Boden (1994) presents a variety of theoretical views and historical-social approaches, each with implications on how to define, study, and understand creativity. See especially the articles by Perkins, Gardner, Martindale, and Eysenck. Different approaches are also found in Smith, Ward, and Finke (1995). For a broad perspective that encompasses behavioral, cognitive, neurological, psychoanalytic, and other viewpoints, see Martindale (1995). A neurophysiological approach is taken by Eysenck (1995). The rarity of creativity is discussed in Walberg & Arian (1999). For different kinds of creativity, see Gardner (1983). For a discussion of "everyday creativity," and its contrasts with "higher-order creativity," see Richards (1999a). A rare instance of placing creativity in the context of art history is the text by Martindale (1999). For a discussion of creative institutions, businesses, and organizations, see Haefele (1962) and Nystrom (1979). The major theorists and prominent researchers in the field of creativity, including Freud, Jung, and others, are covered in a two-volume collection (Runco & Spitzer, 1999; see also Runco & Albert, 1990; Sternberg, 1988). Others are found in Sternberg and Dees (2001), which also presents diverse views on creativity within the context of case studies. Six are equally divided between scientists (Darwin, Pauling, Young) and artists (Irving, Donaldson, and Monet). For the general reader, the journal *Generations*, published by the American Society on Aging, frequently contains brief articles about aging centered around cognition, occasionally creativity, and sometimes art. Particularly suitable for this book is a special issue edited by Kastenbaum (1991b), *Creativity in late life* (see also Kastenbaum, 1991a). For a general discussion of the tie between creativity and aging, see Ager et al. (1981–82).

Old age, too, has its share of definitional ambiguities, depending on who defines it. AARP (formerly the American Association of Retired Persons), a major organization for seniors, sets membership at 50, but was once 55. The cut-off point for Social Security eligibility was 65, but moved to 67 recently and may be set later in the future, although eligibility for those who retire early is still 62. In art, artists have depicted the number and stages of life in a variety of ways; and in art history, the old-age style (Part III) is defined as starting at around 60.

A starker parallel, besides shared definitional ambiguities, characterizes the study of old age and late-life creativity: both refer to either decline or continuation (Part II), and give more weight to the former. Like the sensory and other losses attendant upon aging, creativity also retreats with age: maximized in youth, its span is truncated, fading, and ending well before old age (Chapters 3 and 4).

Countering the decline model of creativity, comparable to successful aging, is the continuity perspective. This model, with its roots in the late 19th century studies of Galton on genius (Dacey, 1999), emphasizes the stability and persistence of late-life creativity. "It should not be assumed that an individual's creativity will stop when he/she becomes a resident of a nursing home" (Syclox, 1983, p. 27). For noninstitutionalized adults 50–90 years, creativity test scores increased after participating in a 4-month quality of life program, compared to a control group not in the program (Goff, 1993). Clinical approaches are also optimistic about late-life creativity (Corsa, 1973). The old can be creative, Pruyser (1987) contended, because they are "survivors" prepared for the hardships of life, capable of sublimating losses, less burdened by ambition, and with time to make up for the lost opportunities of youth. Old age as a period when creativity is enhanced was promoted by Taylor and Sacks (1981) who see it emerging from the elderly's storehouse of accumulated experiences, increased maturity, and freedom for self-expression.

Supporting a continuity model, too, are denotational and consensual criteria of creativity: certain individuals (a Leonardo) and some works (a *Mona Lisa*) are creative if these are the opinions of experts and accepted by others. Accordingly, a large number of 60-, 70-, and 80-year-olds demonstrated late-life creativity (Lehman, 1953; Nelson, 1928; Stern, 1967). Benjamin Franklin invented bifocal glasses at 78, Galileo made his famous astronomical discovery at 74, and Lamarck wrote his major work at 78.

In psychology, Fechner completed his influential opus, *Vorchule der Aesthetik*, at 75, while the first psychology texts were written by Ward at age 66, Wundt at 68, and Brentano at 69. Freud is a particularly apt example. He continued to make creative contributions until his death at 83, including *Ego and Id*, written at age 67, despite suffering from cancer for the last 16 years of his life.

Artists are well-known exemplars of late-life creative achievement. In music, Wagner wrote several of his *Ring* operas (e.g., *Siegfried*) between his 60th and 70th years; Verdi's *Otello* and *Falstaff* were completed at 74 and 80, respectively, and he composed additional operas at 85; and grand operas were written by Meyerbeer at 73 and Rossini at 72.

Among authors, Blake illustrated the *Book of Job* at 69 and G. B. Shaw wrote plays in his 90s. Major works were written by Thomas Hardy at 69, Victor Hugo at 74, Alfred Tennyson at 71, and Samuel Johnson at 68. Cervantes wrote his *Quixote* in his late sixties and Goethe produced the second part of *Faust* at 80. Carl Sandburg penned *Remembrance Rock* at 70 and Sophocles wrote *Oedipus Rex* at 75 and *Oeidpus At Colonus* at 89. James Michener, who died in his 80s, produced 10 books in the last 4 years of his life. The philosopher Santayana wrote a first novel, *The Last Puritan* at age 72. Sherrington, the noted physiologist, turned to writing philosophy at 75.

In the other arts, Giotto created masterworks of architecture at 70. Martha Graham choreographed innovative dances in her 90s, Kurosawa directed three films in his early 80s, and Maurolyuas invented the camera obscura at 81.

Most Nobel Laureates continued to make notable contributions up to the age of their deaths (Pribic, 1990). Of the 84 Nobel Prize winners in literature between 1901–88, 45% were over 65 years old and 61% were over 60 at the time of their award. The oldest was Bertrand Russell, who received a Nobel Prize in 1950 at age 98. (The youngest was Rudyard Kipling, who received his in 1907 at age 41.) Other elderly awardees include G. B. Shaw (who received the Prize at 94), Winston Churchill (91), Anatole France (88), I. B. Singer (86), Herman Hesse and Francois Mauriac (both at 85), Samuel Bechett (84), Henri-Louis Bergson and Andre Gide (both at 82), T. S. Eliot (77), and Jean-Paul Sartre (75).

The *Guinness Book* (Young, 1997) of records has a special section on achievements of the old. In the arts alone, actors with the longest careers include Helen Hayes (78 working years) and Lillian Gish (75 years). Jessia Tandy, at 80, was the oldest winner of an Oscar. The oldest screen performer, the Canadian Jeanne Louise Calmet, was still working at age 105, while a German, Bois, had the longest screen career (83 years). George Cukar made his final movie at 81, while a Dutchman, Ivens, at age 89, holds the record for the oldest director. *King* Vidor directed film for 67 years. In music, the Romanian pianist Delavrancea gave his last public recital at age 103, and the Polish pianist Mieczyslow Horszowski holds the record for the longest international career: 60 years. The oldest active musician, Jennie Newhouse, was an organist for 66 years. Ukrania Reizen sang major roles in opera at 90, while the Japanese singer Tioyotake sang for 81 years. The oldest author, the American Sarah Louise Delaney, published her second book at age 105 and the ballroom dancer Sylvest retired at 94.

Many old painters produced great works up to their deaths. Titian completed his masterpiece *The Battle of Lepanto* and began one of the most famous paintings in the world, *The Descent From the Cross*, in his 90s. Mantagna, Bellini, and Ingres produced great art between ages 70 and 76. Michelangelo worked on his *Last Judgment* between ages 59 and 66 and then did architecture between 67 and 89; he also wrote sonnets between ages 65 and 75. Other artists who enjoyed "a great late phase," include Braque, Matisse, and Picasso; the latter is reported to have said, "As one grows older, art and life become one and the same" (Richardson, 1997, p. 31). Goethe at age 79 claimed, "One must be young to do great things"—and then changed his mind (Nelson, 1928, p. 303). A collection of late works of art is felicitously titled, *I am still learning: Late works by masters* (Feldman, 1992).

Some artists achieved great and unexpected fame late in life (Cohen, 1997; Rodeheaver et al. 1998; Root-Bernstein, 1999, see especially p. 462; Steinberg, 2001), the most well known of whom is Grandma Moses (Mary Robertson). She took up embroidery at age 67 to cope with the loss of her husband, began painting at 78 when arthritis made detail work too difficult, and continued to work at her easel until her death at age 101. Other individuals turned to art late in life after working in other fields and achieved renown. William Edmonson, the first black artist to have a solo exhibit at the New York Museum of Modern Art, began sculpting in his mid-60s after he lost his job as a janitor in a hospital. A sharp career change also describes the life of Sister Gertrude Morgan. She took up painting at 56 while running an orphanage, began painting full time a decade later, and had shows in museums when she was in her 70s to 80s. Bill Traylor, born a slave, achieved national prominence in 1982 with a show at the Corcoran Gallery in Washington of some 1500 works produced between 1939 and 1942 (he died in 1947; A self-taught folk artist, 1990).

Not "late bloomers" but "latecomers" were van Gogh and Cézanne, who took up painting in their 30s and 40s. Louise Nevelson became a sculptor in her 40s, and Edith Wharton began writing in her 40s. In the category of "persisters" are Henri Poincore in mathematics, Richard

Feynman in physics, and Linus Pauling in physical chemistry, to name
a few of those who continued to make creative contributions throughout their lives. (Already mentioned were George Bernard Shaw,
Picasso, Verdi, and Martha Graham.) In contrast, "field changers" made
their major contribution in youth and elaborated on it with exemplary
work throughout their adult careers (Root-Bernstein, 1999, p. 467).
Polymaths, like the world-famous composer Borodin who was also a
foremost chemist, were successful in more than one career throughout
their lifetimes.

For these and other elderly luminaries, physical, mental, and other
losses had a minimal impact on their late-life creativity. Mind and
spirit—or as some might prefer to call them, mental abilities and
motivation—are not as prone to deterioration as the senses and the
body. The superior accomplishments of many, listed above, are notable
exceptions to the expectation of losses with aging.

As impressive as lists of aging overachievers may be, they are limited by their anecdotal status. Names of creative old-timers come to the
fore or fall to the wayside because of current fads and the vicissitudes
of popular culture. How many elderly newsmakers from the past are
remembered now? How many will be remembered 25 years from now?
Have most, whether from the past or present, been recognized?
What about the unnamed 60- to 80-year old also-rans who came in
second or third behind the record-holders in the *Guinness* compilation? A more systematic way of defining creativity among the aged uses
productivity.

Systematic productivity counts of outstanding efforts track creativity at different ages. When and how many great novels were written, masterpieces painted, break-through inventions patented,
prize-winning articles published, enduring music composed, awards
won, and so on? A quantitative index of creativity is not as superficial
as it might seem (Part II). Among scientists, there is "a reasonably good
correlation between the eminence of a scientist and his productivity
of papers" (Price, 1963, p. 40). In support, Simonton (1988b) asserts
that "On both empirical and theoretical grounds [creativity and pro-
ductivity] can be used almost interchangeably; total quantity of output
is intertwined with selective quality of output" (p. 409). (See also
Simonton, 1990a, 1994, 1997, 1998, 1999b).

Productivity plotted over time can also be tied to other objective
markers, such as the onset, duration, and cessation of wars, revolutions, and economic depressions, as well as subjective assessments
by experts of individuals' reputations, along with the occurrence of
critical life experiences, such as the onset and duration of illness.

Thus, a simple frequency count of the number of recognized outstanding works produced by individuals reputed to be creative, linked to any number of factors, suggest possible influences on creativity. Of major importance, the productivity of an individual and groups (artists, scientists) can be tracked against age in order to access the course of creativity. Does it decline, revive, or (re-)emerge in old age?

The connection between productivity and creativity is not without its critics (Levy & Langer, 1999; see Chapter 4). Prolific hacks are not necessarily creative and the author of a single work can be creative. Keep in mind, though, that other ways of studying creativity have problems, too.

Consider creativity tests. The young are more familiar with tests and testing than the old, and are accustomed to working fast and under pressure. These advantages could account for high scores on creativity tests. Another worrisome feature of creativity tests is there are so many of them (75 in a recent count; Runco, 1999) and they come in many formats: multiple-choice and open-ended, verbal and perceptual, checklists of activities and personal traits, projective tests, biographical inventories, measures of divergent thinking, and domain-specific questions about poetry, music, and so on. The large number and diversity suggests that creativity is too complex for one test to capture, or more critically, each test is an attempt to overcome the deficiencies of others (Hocevar, 1981).

In any event, creativity scores of test-wise students during the school years are poor substitutes for more direct evidence of creativity, that is, real-world accomplishments sustained over a lifetime. However, there are relatively few older individuals whom lives are marked by sustained creativity, but many for whom test scores of creativity can be easily obtained. As for the historical record, past figures of creative note are obviously not available for research, and archival records and biographies, if they exist, are often incomplete and unreliable. For creative people still alive, in-depth interviews and entries in personal journals, diaries, and autobiographies (Kastenbaum, 1992) are prone to interviewer bias, self-serving attitudes, and incomplete memories.

Problems, too, plague strictly controlled experimental studies of creativity. Demands for speed and accuracy favor the young who are "test wise" from classroom experiences with multiple-choice and standardized exams that impose strict time limits and demand one required answer. College students are also familiar with research materials and laboratory procedures from participating in studies as volunteers, often as part of class requirements, and have taken science classes as well as read about research in their textbooks. Younger research participants therefore find it easier than the elderly to accommodate themselves to the unnatural requirements of unusual laboratory environments with unreal materials. The elderly's unease over the strange constraints imposed on them are amplified by poor eyesight, hearing deficiencies, reduced motoric coordination, and arthritic pains (Chapters 15 and 17).

Thus, losses in late-life creativity in a range of studies may be more apparent than real. Poor performance among the old may also be due to reasons other than age. Consider the differences usually found in cross-sectional studies that compare 20- and 60-year-olds to the disadvantage of the older group. However, generational differences in educational, socioeconomic, and other historical experiences of 20- and 60-year-olds affect their performance on creativity measures, not age, per se. Thus, the older group, because of having served in the military during a war and lived through an economic depression, had fewer opportunities to participate in creative activities when young. These interpretive problems are minimized in longitudinal investigations, like those of Terman and Erikson, discussed earlier, where the same 20-year-olds were tested until they were 60 and older, which means they were influenced by the same external forces over time. Yet longitudinal research is rarely done because it is difficult and expensive to track the same people over many years (assuming they are still alive).

Questions about research design, methodology, tests, and qualitative research are enormously difficult in general, but are exacerbated when applied to aging creative individuals. They may not, understandably, be willing to participate in research because they are reluctant to give up their limited time and diminished energy for research that is of no obvious benefit to them. Furthermore, many years have to pass after creative persons' deaths in order to assess and affirm their status.

Every approach to late-life creativity, ranging from the laboratory to the historical record, whether cross-sectional or longitudinal, or relying on tests or interviews, therefore poses dilemmas. These uncertainties, along with the definitional issues raised earlier about creativity and aging in general, temper research on late-life creativity. Consequently, late-life creativity has not taken center stage, and when it has, decline rules with its large number of confirmatory studies. The paucity of data on late-life creativity is taken as confirming the prevailing view of aging as a time of loss. Thus, the youthfulness of creativity, not its agelessness, has attracted most attention (Chapter 3). Many have concluded, as if it were common knowledge or a foregone fact, that creativity is minimal or absent by age 60 or 65, and if it persists, it does so for only a very fortunate few, and moreover, it is weaker than in youth. The mode of creative expression, furthermore, is believed to be set in youth and not likely to change or emerge in old age. But the evidence, as noted, is open to reinterpretation, and as the chapters that follow will show, the decline model is countered by other research, little as it is, much of which is new, not widely reported, or known.

Obstacles to the study of late-life creativity should not derail or delay optimism about the future. Definitional and other difficulties can be set aside, at least temporarily, so that investigations can be initiated and proceed, albeit cautiously, and then revisited and revised later when researchers are better armed with more knowledge. An overpreoccupation with methodological and other stubborn problems, if belabored, stymies the search for solutions. The creativity of old people merits serious consideration, not only for its newsworthiness or as a personal role model, but also because it exemplifies a positive aspect of human growth. For practical reasons, too, the affirmation of late-life creativity supports programs that encourage the elderly to engage in activities in the arts and other areas that further its expression. The fact of late-life creativity is quite real, as attested to by the large number of elderly achievers; creativity in old age is not a shadowy, minor, or peripheral phenomenon.

Unfortunately, research on late-life creativity is sparse, especially the empirical kind, a term used interchangeably with "scientific," "objective," and "factual" (Balkema, 1986). Empirical research is characterized by careful observation, controlled or carefully defined conditions, and quantitative measures. Contrasting but potentially complementary approaches include clinical claims about late-life creativity, narrative discussions among scholars in the humanities about aging, and personal accounts by aging creative individuals, such as artists, along with anecdotal and archival materials, and speculative accounts by and about famous elderly people.

The small number of empirical studies of late-life creativity is startling when compared to the large number that focus on creativity in the first quarter of life, on aging and creativity in general, and the less sanguine aspects of development (Bornstein, 1984; Cohen, 1986; Kogan, 1973; Simonton, 1990b). Butler (1973) justifiably complained that, "Fresh bursts of creativity are rarely given recognition in later life" (pp. 98–99). The reason, the Romaniuks (1981) charged, is investigations are mainly "directed to creativity in children, adolescents, and young adults." Consequently, "relatively less research energy has been invested in creativity in adulthood and especially old age." The result, the Romaniuks concluded, is "an unwarranted pessimistic view of creativity in late adulthood" (all quotes are from pp. 366–367). Baltes and his colleagues (1990) agreed. "[T]he empirical harvest is relatively small [for the] truly superior performances of older adults" (p. 64). For Kastenbaum (1991b), the paucity of research on late-life creativity arises from the overemphasis on indices that record the amount of work, rather than the more qualitative aspects of old age, such as its "creative spirit" (pp. 5–6). Consider, too, that late-life creativity occurs relatively infrequently, just as there are only a few old wise people (Clayton, 1975). Research on creativity among the aged is also hampered by problems inherent in the study of old age in general. Resisting its study, too, are the difficulties inherent in the study of creativity at any age (Balin, 1994; Csikszentmihalyi, 1991; Romaniuk & Romaniuk, 1981; Simonton, 1999a). The status of the topic is indirectly reflected in the many texts and handbooks on aging and creativity, few of which include a section on late-life creativity, especially in the arts where it is compellingly illustrated (cf. Simonton, 1988a, Simonton, 1984a, 1988b, 1994, 1997).

Impeding research, too, are societal prejudices against the aged which make it difficult for old people to be creative or to be recognized as such. "Old age is not for sissies…. Those who continue to renew their creativity in later life [need to find] the courage and independence to resist the dominant climate of [negative] expectations" (Kastenbaum, 1991c, pp. 5–6). The elderly also accept societal prejudices against late-life creativity. "Earlier pessimistic conclusions about declining creativity with age," noted Moody (1982a, p. 246), although mistaken, lead to "stereotypes and expectations that discourage creativity in later life…. Older people have been victims of this self-fulfilling prophecy of decline." Moody was hopeful of change, though. "As [social] conditions change, that tendency could become reversed." Aiding that goal, he observed, were "cultural programs for older people."

LOOKING AHEAD

There are many approaches to creativity: scholarly, clinical, practical, personal, and empirical. This book emphasizes the empirical. Similarly, there are several kinds of creativity: scientific, scholarly, and artistic. This book focuses on the latter, for reasons discussed in Chapter 2. In the arts, this book highlights painters and paintings (Parts III–V). Thus, an empirical approach in the framework of the arts presents arguments and evidence on late-life creativity among aging artists and late-life works, with particular attention paid to old painters and late paintings. In addition, elderly nonartists and their responses to paintings and their engagement with other forms of the arts are examined (Part V). The larger implications of these chapters on artists and art on the course of creativity, aging, late art, aging artists, and arts activities for the elderly are discussed in Chapter 17, along with the practical relevance of the arts to the aging public, the contributions of the arts and humanities to science, and interdisciplinarity.

There are many determinants of creativity: psychological, inter-personal, sociological, and social-cultural. These pages concentrate on the psychological. Within psychology, creativity is about personality, motivation, and cognition. This book promotes the cognitive aspects of creativity: the roles of perception, thinking, judgment, conceptualiza-tion, imagination, imagery, and problem-solving (Jaquish & Ripple, 1981; Gruber, 1974). Thus, creativity-as-cognition is about problem-finding, -defining, -identifying, -discovering, -expressing, -posing, -representing, -translating, -integrating, and -synthesizing (Richards, 1999a; Runco & Dow, 1999). Thus, artists were observed to literally grasp objects in order to select, group, and manipulate arrangements that not only explored their aesthetic properties but also sensitized them to the artistic problems they posed (Getzels & Csikzentmihalyi, 1970). Physical "hands-on" explorations restructured and integrated ongoing ideas from which creative solutions emerged.

Stated more broadly, creative people perceive more, learn faster, think better, make quicker decisions, and engage in a larger variety of unusual (divergent) modes of thought. They also make more connec-tions, have richer associations, and produce highly vivid images (Lindauer, 1983). Creativity therefore represents "the best" kinds of cognitions, and late-life creativity reveals the cognitive capabilities of late-life development.

There are two ways to unravel the cognitive aspects of creativity: "from above" or "from below" (Lindauer, 1984d) The first approach (Parts III–IV) begins with aging creative individuals, in the present case creative old artists and examples of their extraordinary achievements. Knowledge about late-life creativity gleaned from exemplary older individuals "moves down," so to speak, to "everyday creativity" as it occurs among so-called "normal" old people. Alternatively, starting "from below," the focus is on ordinary older citizens who are engaged in creative artistic events and activities, like looking at a painting or attending an art museum (Part V). The study of cognitive abilities needed for creative activities "moves up," so to speak, and helps understand individuals who are recognized as creative. Approaches "from above" (starting with old artists and late art) and "from below" (beginning with aging nonartists engaged with art and creative activities in the arts) eventually converge in elucidating general issues of cognitive development and aging.

This argument presumes that elderly nonartists who participate in a creative artistic activity depend on cognitive and other processes that are like those in old artists who create art (Richards, 1999a). Both nonartists and artists try to figure out what a work might or should mean: they interpret, judge, imagine, evaluate, and make decisions. With few or no clear guidelines, both nonartists and artists search their memories, associations, ideas, and other mental resources; reflexive, routine, or habitual responses will not do. Artists and nonartists purposefully search for relevant connections, discard extraneous paths, and organize what remains. Ideally, there comes a point where a work of art in progress (by an artist) or under observation (by a nonartist) is better understood than it was before these cognitive explorations began. If and when this process succeeds, both artists and nonartists experience an insight, a breakthrough, an aesthetic "aha," a creative moment.

By this reasoning, creativity is not a unique phenomenon, the product of genius or a special talent, but related to everyday cognitions. This radical idea was developed by Bink and Marsh (2000; quotes are from pp. 59–60, 75). Creativity is a "product of more general human abilities that might be possessed by many." Consequently, "a creative achievement can occur through...mundane means." In other words, a "common underlying cognitive processes [can] give rise to both more creative and less creative products." Everyone, not only "the creator faces the problem of deciding what information is most relevant in moving from a blank slate toward an appropriate end state...by generating relevant information, synthesizing that information, and selecting from among that information." All cognitions, creative or not, include "working memory, capacity, speed of retrieval, perceptual fluency, activation of relevant concepts and inhibition of irrelevant ones, recollective ability, inspection of memories, and a host of other processes that are used in everyday cognition." Why are some cognitions creative and others ordinary? The answer for Bink and Marsh: "Individual variation in the same cognitive processes."

If creativity is like other cognitive abilities, it diminishes with age for some, is left unaffected for others, and for a few, emerges refreshed or for the first time late in life, like wisdom. A certain number of older individuals work around barriers to creativity, others cannot, while some require help. Individuals differ in their reactions to aging and how it affects them, and this is as true for late-life creativity as it is for any other trait, process, or behavior.

This book champions late-life creativity but also recognizes that it is neither guaranteed to occur nor stay, just as youthful creativity does not last indefinitely. Precocious children fail to fulfill their youthful promise, individuals once hailed for their lifetime achievements become footnotes to history, and geniuses from bygone eras are now forgotten and their achievements are no longer celebrated. "An artist renowned in his or her lifetime may be obliterated from history books" (Stephens, 1998, p. 38).

At the same time, it is important to state that creativity and aging are not contradictions. To hold that late-life creativity is non-existent or limited, hampers thinking and research on old age, cognitive development, and late-life potential in general; discourages the elderly's participation in creative activities; and makes it difficult for legislative bodies and communities to promote creativity-enhancing options for the elderly. Thus, for scientific, practical, and personal reasons, as well as on idealistic, ideological, and humanistic grounds, this book takes an optimistic stance on the persistence of creativity into old age. The reader, at the end of the book, will hopefully agree.

CHAPTER 2

Old Age and Old Artists

Old age hath yet his honour and his toil;
Death closes all; but something ere the end,
Some work of novel note, may yet be done,
Not unbecoming men that strove with Gods.
Tennyson, *Ulysses*

The arts have great appeal: they require little or no expertise to be accepted or rejected, appreciated or despised, unlike scientific and technological accomplishments. A glance at a painting, the sound of the opening chords of a symphony, and the sight of the first few lines of a novel instantly arouse interest or indifference. Artists, too, are attractive: they are adept at translating abstract ideas into concrete images, illustrating broad generalizations with dramatic instances, softening polarities with subtle distinctions, conveying personal values, anticipating major changes in society, and revealing life's complexities—and in dramatic and compelling ways. The public's fascination with a Zola, Mozart, Rembrandt, and van Gogh were translated into popular biographies, novels, and films (Armour, 1980; Walker, 1996).

Artists and the arts, including "the literary arts," are intimately related to creativity, whether art is broadly or narrowly defined as referring to all forms or to paintings alone (Martindale, 1999, p. 119). Indeed, "art," "artist," and "creativity" are often used interchangeably (Lindauer, 1975, 1983; Winner, 1982).

Only in the arts are the criteria for creativity unquestionable; only in the arts is it possible to assume that an outstanding person is, by definition, a creative one.... No one, for example, would question Shakespeare's or Beethoven's creativity.... [G]enius in the arts is universally considered to be a clear manifestation of creativity. (Rothenburg & Greenberg, 1974, p. xvi; *cf.* Rodeheaver, Emmons, & Powers, 1998, p. 203)

Art and artists are particularly visible exemplars of creativity, especially when reputations endure for centuries (Da Vinci's *Mona Lisa*). Moreover, old artists and the art of their old age bear on late-life creativity, aging, and the course of development.

OLD AGE AND THE ARTS

The elderly engage in a wide range of artistic activities (Cohen, 1997). The elderly in Japan were known for flower-arranging, calligraphy, the tea ceremony, and practicing the martial arts; the story-tellers in African and American Indian societies were the elders; and retired government officials in China engaged in calligraphy and landscape painting. In the United States, the government's official support for arts-for-the-elderly began with a 1976 conference whose stated intent was

> to focus national attention on the role of the arts in developing a more positive image of aging and the aged in our society [and to highlight] the fact that older Americans are an important new constituency for expanding arts programs, as audience, as contributors of time, skills, knowledge and other resources, and as students, teachers and creators. (Sunderland, Traylor, & Peter Smith Associates, 1976–77, from an unnumbered page in the Preface)

The conference was attended by over 200 leaders from arts and aging agencies in 31 states, the District of Columbia, and Canada. They presented 130 reports on a variety of topics related to the aged: increasing access to the arts; building an arts/aging alliance; expanding cultural services, including museum visits; intergenerational differences in arts audience; reaching the institutionalized old; and papers on specific areas of the arts (theater, art, music, and the visual arts) in relation to the aged. The conference was a success in that the arts have a "popular [and] a solid niche in the aging network [which has resulted in an] explosion of late-life involvement in the arts by older people" (Moody, 1982b, pp. 245–246). (See also United States Congress, House Select Committee on Aging, Subcommittee on Human Services, 1980; Useem, 1976; Sunderland, 1982. For a wide-ranging discussion of the role of the government in setting public policies on the arts, see Moody, 1982a, especially footnote #38, p. 300. For the government's role in supporting the arts and how this might be accomplished, see Malcahy et al., 1980.)

On a more personal level, the arts give meaning to the late years of life (Polansky, 1973, see especially p. 111). A novel, poem, and a play about an old person or growing old that expresses optimism about and acceptance of aging can be uplifting and cathartic. Quotations, too, articulately capture thoughts and feeling about aging, especially those from Shakespeare and other notable literary figures, and the Bible. Brief and pithy, and highly literate, quotations transform vague thoughts and muted feelings into insightful capsules about aging. The specialized collection by Sampson and Sampson (1985) arranged quotations for every year of life, from birth to 90, with about half falling into the last third of life. The compilation by Booth (1992) included quotations about old age from poets, essayists, and writers that illuminate "losses and fears, lamentations, cures and consolations, and celebrations," and the mixture of joys and sorrows that accompany aging (from the table of contents). A few examples follow:

> The dry branch burns more fiercely than the green. (Elder Olson, written in his mid 70s)
> I am more myself than ever before. (Mary Sarton at 60)
> We are all happier in many ways when we are old than when we were young. The young sow wild oats, the old grow sage. (Winston Churchill)

The collection also contains about 30 portraits and photographs of aging notables (Walt Whitman, Robert Frost, Georgia O'Keefe, G. B. Shaw, W. H. Auden, Albert Sweitzer) and art that depicts old people (by A. Durer, Corot, van Leydem, Benton, Luks, A. Wyeth, Rembrandt, Ben Shahn, Kollwitz, and Renoir).

Attitudes towards aging are also succinctly summarized in Tin Pan Alley songs (Chudacoff, 1992, quotes that follow are from pp. 151, 153–154, 155–156; see also Cohen & Kruschwitz, 1990). The lyrics from the 1920s and 1930s, despite their triteness, sentimentality, and romanticism, suggest that Americans "accepted old age as a separate, identifiable, age-bounded time of life, fraught with special needs and dilemmas," such as "the distress and disability accompanying the aging process," "fear and degradation of the elderly," and "sympathetic and empathic attitudes toward old age." More recent popular music has dropped many of these earlier themes and added new and opposing ones. "To be young, to feel young, and to act young" were the predominating motifs in the 1990s.

LITERATURE AND OLD AGE

Literature is an especially valuable font of information about old age, Cole (1992a) argued, because it portrays and interprets commonplace thoughts and feelings about aging that are experienced in everyday life. Fictionalized accounts of aging have an advantage over scientific studies of late life, he continued, because they are less formal, logical, and rational. Literature therefore supplements and enriches abstract conclusions based on tests and other quantitative measures, and the ordinary materials and artificial conditions of the laboratory.

Gerontologists, though, have only recently become interested in literary works on aging, which Wyatt-Brown (1993) calls "Literary gerontology" (p. 3). Literature, she claimed, adds a welcome positive outlook to aging that balances more negative positions on old age. Achenbaum (1989) agreed, and considers literature a vital part of "humanistic gerontology" for generating "a portrait of aging in a cultural and historical perspective" (p. xxiii). In support, too, Yanke and Eastman (1995, p. ix; see also Yanke, 1994) insist that literature offers a unique in-depth analysis and case study of aging. "Literary works... illuminate the problems of older people [for they] go beyond the discipline of gerontology... to convey the experiences of aging... to show aging from the inside as the interpreted perceptions and feelings of individual lives [and the] concerns of art, not science." To buttress his argument, Yanke and Eastman's (1995) research manual contained 360 references to studies of novels, short stories, plays, and poems about aging. These fell under 44 topics and were grouped into five major areas related to aging: society, relationships, physical health, psychological responses, and the search for meaning.

Studies of the literature on old age were indexed in Eyben's (1989) nine-page bibliography, beginning with the Bible and Greek and Roman sources. Ancient "writers' perceptions about aging... make important contributions to gerontology" by educating people about old age (quote is from the unpaginated abstract). Falkner and Luce (1989) put the case more strongly. The meaning of old age in contemporary society, they claim, can "only be appreciated by an exhaustive comparison [of Greek and Latin literature with that of] other societies and times" (p. vii).

Discussions of literature and old age are numerous (Achenbaum, 1989; Bagnell & Soper, 1989; Kastenbaum, 1989, 1994a; Lyell, 1980; Woodward, 1994; Woodward & Schwartz, 1986; Wyatt-Brown, 1988, 1990, 1992). The ambitious work by Wyatt-Brown and Rossen (1993) focuses on aging women's creativity from the perspectives of literary criticism, biographical theory, cultural anthropology, and history; emphasizes the careers of Sarton, Bogen, Swift, Woolf, Bowen, among others; and arrives at a generally positive view of aging. A few examples give a sense of the field.

Coffman (1934) surveyed changing views on aging, beginning with the poet Horace (20 BC), moving forward to Chaucer (14th century), and ending with the Medieval period. Writers' observations in these periods, Coffman concluded, reflected the moods and intellectual spirit of their times. Covering a longer time span was Charles (1977), who began with the Bible and ended with 20th century fiction, concluding that themes about old age reflected the societal values of those times as well as reinforced them. Charles also noted that portrayals of the elderly in the 19th and 20th centuries were more positive than earlier periods, which is in agreement with others, although old women were not well represented. The Medieval period was the focus of Goodich's (1989) study of the writings of saints on aging. Parallels with contemporary views of aging were again found. Literary studies of aging also encompass fairy tales, which generally present a positive picture of the elderly (except for wicked old witches, of course, as well as poverty), according to DeAngelis's (1989) summary. Fairy tales largely emphasized "the psychological challenges of later life," including "self-reformation, transcendence, worldly wisdom, emancipated innocence, and mediation with the supernatural working on behalf of the next generation" (p. 40).

Authors, too, articulate their personal reactions to aging. Particularly good examples are the journals and diaries of five well-known professional writers between ages 59 and 85 (Elizabeth Vining, Mary Sarton, Doris Grumback, Florida Scott-Maxwell, and Alan Olmstead), who received a close reading from Berman (1994) within the framework of several psychodynamically oriented theories of personal development (by Jung, Erikson, and Levinson). On the key question, "What is it like to be an older person?" no single answer emerged, not surprisingly. Each writer, Berman reported, faced different problems of aging, attempted to solve them in a variety ways, and arrived at a host of solutions. Thus, the authors reported a range of difficulties faced with increasing age, how they were met, and whether they were overcome, as well as commented on constancy and changes in styles, technique, and subject matter as they aged. No frequencies to their answers were assigned. Unknown are the kinds of reactions that predominated or whether the single man (Olmstead) differed from the four women. Autobiographical material from aging writers are a rich source of information that have not been systematically studied with some attention to quantification.

Literature has also been examined in conjunction with the visual arts. A good example is Covey's (1989) survey of the elderly in the literature and art of the Middle Ages to the 18th century (see also Covey, 1991). Literature and the visual arts, he concluded, consistently disclosed three major themes about aging: ambivalence, reactions to physical decline, and judgments about the appropriateness of age-related behavior. In an even more ambitious study, Achenbaum and Kusnerz (1978, 1982) compared quotations from literature and musical lyrics with engravings, etchings, Currier & Ives prints, drawings, lithographs, and news photos. Achenbaum and Kusnerz noted, as have other studies of trends over time, that societal attitudes about the old were consistently reflected in varied sources. Paralleling the findings of others, Achenbaum and Kusnerz also found that attitudes were initially positive, then became negative, and are apparently becoming more positive. (For a similarly comprehensive case study of aging from a clinical perspective that emphasizes individual works of art and literature, including poetry, as well as popular art and photos, see Bertman, 1989.)

PAINTINGS, THE VISUAL ARTS, AND OLD AGE

Portrayals of the elderly in the visual arts have received more attention than literature because of several advantages (Lindauer, 1984e, 1987b). A painting is concrete, as is literature, but it is also succinct: its "story" or meaning can be grasped in a glance, in less than a second. Only quotations, epigrams, and haiku poetry, the opening bars of a symphony, and musical lyrics approach this immediacy. Thus, paintings depicting the aged can be quickly comprehended, as can comparisons between works by young and old artists. In contrast, a novel or play about an old person requires a fairly long time to unfold, many characters have to be kept in mind together with their motives, often obscure, along with sub plots whose meanings are not immediately obvious. Complex information about a central aging figure is easily forgotten or distorted; and interruptions, delays, and disruptions that occur over an extended reading period increase the likelihood of memory lapses. Similar errors characterize other art forms that unfold over time, like theater, which also requires a great deal of information to be stored and then retrieved. Similarly, the ability to detect changes between a composer's youthful and later works depend on listeners' attention over a long span of time. In contrast, the speed with which paintings capture our attention allows for an immediate impression, and a flow of ideas, thoughts, and images that are less subject to distractions, interruptions, and losses in information.

Paintings that depict old age are therefore good candidates for aging research (Berg & Gadow, 1978). "It would be difficult to find a major [painter]" Kauppinen and McKee (1988) asserted, "who has not given at least some attention to [old age]" (p. 88). They analyzed several hundred images of old age in paintings (Kauppinen, 1991; McKee & Kauppinen, 1987) and observed that they did not conform to stereotypes about aging. The life styles of the aged were more active than withdrawn, there was more growth than decline in wisdom and mental development, and on a slightly less positive note, negative intergenerational exchanges between young and old were equal to positive ones. In addition, Kauppinen (1991) examined physical, social, and psychological stereotypes of old age in over 400 paintings by nearly 200 Western artists, concentrating primarily on four themes: the abuse of power, uncleanliness, incompetency, and obsolescence. Most works, she again found, were positive or complimentary, not ageistic. For example, the old were depicted as sensuous and interested in sex. Only a small number of works, like Leonardo's caricatures of physical deterioration and Goya's ugly representations of the elderly, could be called negative.

Kauppinen (1991) advised caution, though, in using art for scientific purposes: "Artists are not illustrators [nor theorists] of social and psychological theories" (p. 231). She warned that the diversity of images in art are quite subtle and not easily categorized. Kauppinen's wariness about using art for scientific purposes might be allayed if a panel of art historians confirmed her interpretations. Helpful, too, would be frequency counts of the themes represented in art, distinguishing between works by young and old artists whose images of old age may differ, and contrasting portrayals of aging in different cultures and time periods.

Not all of the art on aging is positive. Jacques (1980) examined depictions of old people in the art of different cultures, a subject that has "an ancient and honorable tradition" (quotes here and elsewhere are from pp. xviii, 117, and 119). Chinese scroll paintings, for example, depicted the aged as both respected and feared because of their authority. Similarly, in early Western art, there was a mixture of respect and fear in portrayals of God and the Fathers of the Church, who were invariably shown as old. With Byzantine art, Jacques observed a shift to highly negative portrayals, with old people illustrating the terribleness of aging: "menacing, admonitory, adamantine." In the Renaissance, though, artists like Michelangelo, Leonardo, Titian, and Tintorello treated age as honorable. However, Jacques interpreted Rembrandt's self-portraits as an old man as unfavorable to the elderly. Works by Rembrandt, which Jacques believed were keen observations of old age, include ...*In An Armchair,* ...*In Fanciful Attire,* ...*Wearing a Linen Head-Band,* ...*With a Gold Chain,* ...*With a Stick,* ...*Wearing Fur Cap,* ...*In a Cafe, Study of the Head of an Old Man, The Apostle Paul,* and simply, *An Old Man.* The artist Ghirlandaios' *An Old Man and His Grandson* was noted as particularly positive towards aging as was Titian's *Votive Portrait of the Vendramin Family.* Jacques' own drawings of aged residents in two nursing homes represented "pathos and tragedy," not unexpectedly, since he called his experience in nursing homes "Dante-esque travels among people condemned through no sin of their own to modern Infernos and Purgatorios while still on earth."

More neutral are artists' depictions of "the ages (or stages) of man," essentially a longitudinal conception of aging (Hoffman, 1965; Sears & Feldman, 1973). Especially detailed was Cole's (1992) analysis of the stages of life in the major serial works of three artists: Thomas Cole's *The Season of Life: Old Age* (1842); Guiguin's *Where Do We Come From? What are We? Where Are We Going* (1897); and Jasper Johns' *The Seasons: Winter* (1986). In general, artists have depicted three or four stages of life: childhood, youth, maturity, and old age. Some, though, portrayed as many as 12. Irrespective of the number, the stages generally suggest two paths taken with age: linear, moving forward; or cyclical, with individuals returning to earlier periods as they aged. The stages also differ in duration and complexity, which hints at similarities and differences between the old and the young, and the relative importance of each stage to one other.

The diversity of ways in which stages are depicted, as well as their varying number, suggest that growing old has different meanings over time, at least as artists see it. This rich source of information about aging has received little systematic study, starting with a review of the different stages of life portrayed by various artists throughout history, followed by a division into those with positive and negative qualities, and concluding with parallels to contemporary conceptions of aging.

The task is complex, but a model of what can be done is Kastenbaum's (1994b) analysis of nearly 500 images of Greeks, Romans, and saints in paintings over 19 centuries, along with familiar and popular legends and religious writings at those times. Kastenbaum discerned three models of aging in the paintings and writings. The first is "The Saint," the old man without vanity, vices, or sin, a person who is marginal, outworldly, detached, ascetic, and burdened. The Saint, in response to unusual experiences, developed extraordinary thoughts, ideas, and beliefs. The Saint is a difficult model for old people to emulate, but it is a positive one of aging. A more negative model of aging is "The SOB" (son of a bitch), the wrong-headed old man who unapologetically and aggressively opposes youth and progress. Yet the SOB also has admirable traits, too. He is not a helpless victim, but has a strong sense of self and purpose. The third model is "The Sage," a figure of moderation, equanimity, and balance, the elderly person who transcends personal ambitions and inner turmoil.

The three models, although abstract ideals, contain some essential truths about old age, Kastenbaum claimed; there are aspects of each of them in all old people. Thus, the aging body is imperfect ("The Saint"), the aging spirit is unrequited ("The SOB"), and the aging individual is observant ("The Sage"). None of these types, Kastenbaum emphasized, conceptualize the old as either helpless or incompetent.

Three different models of aging in art were also extracted by Winkler (1992): negative, the old as grotesque, embittered, and despicable (Goya's *Hasta La Muerte* [*Until Death*], 1881–86; and Grosz's *Ecce Homo*, 1919); positive, the elderly as wise, gentle, reflective, and blessed (Ghirlandaio's *An Old man and His Grandson*, c. 1480; and Rembrandt's *The Nature of the Prodigal Son*, c. 1665); and lastly, transcendent, as when the aged face death (Michelangelo's three *Pietas*, c. 1499, 1550–55, and c. 1555–1564); and four works by Kathe Kollwitz (d. 1943), including a self-portrait, undated except for one sketched in 1938).

A somewhat similar set of three categories emerged from a survey of visual images from America's past to the present by Shenk and Achenbaum (1994; see also Achenbaum and Kusnerz, 1978, 1982). They found negative images, where the emphasis was on physical decline and reflected ageism; positive images, characterized by wisdom and successful aging; and "adult," or mixed views, which combined the previous two categories.

The art of old age not only illustrates various conceptions of aging, but also suggests hypotheses regarding the course of aging. Some portrayals may even foreshadow scientific discoveries. "Pictorial masterpieces," Kauppinen (1991) asserted, "anticipated many important observations and concepts of recent gerontological study [including] the tension between active and disengaged life styles in late life, the life review, the integrative understanding [that comes with aging], and [the notion of] intergenerational passing" (p. 217). Other themes that anticipated later empirical inquiry were the association between age and wisdom and the conflicts between younger and older generations. On a more personal level, the amateur art of the institutionalized elderly revealed a "precognition, premonition, or foreshadowing of death" (Zlatin & Nucho, 1983, p. 113). Two rival hypotheses about late-life creativity in art, either decline or stability, are examined in late art and old artists in Part II. First, though, it is important to establish that artists are sufficiently long-lived to demonstrate a late-life creativity in their works. If artists and others are indeed short-lived, it would be impossible to study late-life creativity for there would be few individuals, if any, to study.

THE LONG-LIVEDNESS OF ARTISTS AND
THE STUDY OF LATE-LIFE CREATIVITY

Van Gogh, Correggio, Raphael, and Toulouse-Lautrec had brief lives. In music, Mozart, Schubert, Mendelssohn, and Chopin died before they were 40. The early deaths of these artists, however, is not typical. The short-livedness of a few creative people fosters one of the many myths about creativity, fueled by the selection of a few well-known celebrities. (Other myths are that creative people, especially artists, suffer for their extraordinary talents, and that madness furthers creativity (Becker, 2001; Lindauer, 1994).)

In fact, artists as a rule remain actively engaged into their 70s, 80s, and longer, rarely retiring until very late in life, if ever, unlike aging members of most other occupations (Lindauer, 1993b; see also Chapters 6 and 7). There is considerable evidence for artists' longevity. A popular text of world art (Arnason, 1986) devoted at least one paragraph of text to 200 artists who lived to at least 60, accounting for 68% of those discussed. A survey of 250 major Western artists included 65% who were 60 and older when they died (Held, 1987). Another list of 150 major historical artists of renown lived to at least 60, and when the number was expanded to 225 by additions from a panel of art experts, nearly a third (70, or 31%) lived to 79 or older (Lindauer, 1992, 1993; see also Chapter 13). Declines in abilities were not detected in a sample of 344 older Chinese painters (Silbergeld, 1987). Of 233 recognized masterpieces of art, about two third (76%) were by artists who lived to 60 or more (Chapter 7). A list of the 1,000 most eminent men who ever lived, compiled at the beginning of the 20th century, included 17 artists, and of these, 13 (76%) died when they were over 60 (Cattell, 1903). Exceptional and long-lived artists who produced great paintings in their old age (and the ages at which they died) include Bonnard (80), Degas (83), Goya (82), Hals (86), Matisse (84), Michelangelo (89), Munch (81), Renoir (78), Rodin (77), Tintorello (76), and Titian (92). Numerous special exhibitions of art by old artists have been mounted (Chapter 5). "Artists need never fade away" (Greenberg, 1987, p. 25).

Lehman (1953), the most well-known advocate of the youthful onset and cessation of creativity (Chapters 3 and 4), forcefully insisted that "The vast majority of our great creative thinkers have not died young.... It is simply not true for most creative individuals that 'Whom the gods love die young.'" It is just the opposite. "On the average," Lehman continued, "the creative individual lives as long or even longer than do lesser mortals." Lehman hoped these figures would be "an encouragement to older individuals" (quotes are from pp. 273 and 275).

Lehman held a different view on the years in which creativity *peaks*, as will be shown later.

The number of long-lived artists of renown is therefore appreciable. Artists who remain creative during an extended span of life, and in many cases produce enduring masterpieces, demonstrate that creativity need not decline nor end before old age, that age-related deficiencies can be overcome, and that cognitive and other abilities that undergird artistic talent do not necessarily wane over time. The late-life accomplishments of creative artists are remarkable in light of many living in times when life spans were considerably shorter than today's. Thus, their average life spans, already high, are in a sense even higher when weighed against the short-livedness of the population at large during the early time periods in which most of these artists lived, when poor health was the norm, people generally suffered from a wide range of ailments, and reaching age 60 was considered very old and a remarkable achievement.

The availability of a large pool of long-lived artists therefore makes it possible to track peaks, plateaus, and declines in creativity, if any, and the ages in which they occurred (Chapter 3). The long-term record also prompts other key questions about late-life creativity: Do young creative artists remain creative as they get older, or more or less creative? Does creativity increase over time? What other trends in creativity over time are there besides loss, stability, and growth (Parts II and III)? Do patterns of creativity vary for artists who work in different mediums? These questions assume that the creative profiles of long-lived writers, composers, and other artists, including painters, can be compared at different ages, especially old age.

The inclusion of long-lived old artists and the work of their old age extends the scientific study of aging to art history, musicology, literary criticism, scholarship on the arts in the humanities, reports from aging artists, and late-life art. Contributions from various disciplines supplement, amplify, and corroborate scientific studies that for the most part rely on nonartists who are tested with nonartistic materials. Equally important, converging input from multiple disciplines furthers interdisciplinary study.

THE STUDY OF LATE-LIFE CREATIVITY AND INTERDISCIPLINARITY: THE PROMISE AND THE REALITY

The study of late-life creativity intersects with a number of disciplines: psychology, gerontology, the arts, and the humanities, as well as recreation, leisure, and museum studies. Achenbaum (1989) adds "folklore, myths, magic, rituals, art, diaries, poetry, songs, proverbs, social mores, and legal codes [as well as] medical etiologies and therapies, novels, ethnic stereotypes, jokes, novels and television programs." Art and literature in particular hold a special place in aging studies, Achenbaum maintained, because they "describe and explain continuities and changes in the meanings and experiences of being old and growing older [and] the universals of aging across places and over time...that reflect differences in race, class, or gender and...the nature and dynamics...of ideas about age and aging" (p. xiv). In addition, reports from aging artists and examples of their art, given a critical-historical context by the humanities and framed by scientific facts about old age, support public and governmental agencies charged with establishing and maintaining arts programs for the aged.

The host of disciplines represent different methods and goals. The arts emphasize intuitive inquiries and expressive outcomes. The humanities favor narrative analyses of the individual and highlight the unique. Science depends on the tools of the laboratory to disclose broad generalizations. Multiple directions not only stretch the scope of late-life creativity but also challenge the "two-cultures" conflict between science and the humanities (Snow, 1959; Cornelius, 1964). The study of late-life creativity therefore connects a remarkable range of disciplines.

Unfortunately, the promise of a vigorous interdisciplinary exchange has not been fulfilled. Tellingly, a multivolume and continuing series of handbooks on aging (Binstock & George, 1985, 1990, 1996) omits the arts (one chapter on leisure and aging comes closest). There are, though, miscellaneous age-related topics on demographics, social strata, health, income, economics, work, retirement, housing, family relationships, illness, caregiving, hospitalization, race, ethnicity, and mortality. Revised editions update earlier chapters and add new ones on theories, methods, disabilities, cultural and historical perspectives, gender, politics, social services, law, and ethics. But nothing on late-life creativity, aging artists, old art, and arts activities for the aged. Late-life creativity is absent, too, in specialized texts on aging where its inclusion would be appropriate. For example, a chapter on the practical aspects of cognition applicable to the elderly (Park, 1992), such as remembering to take medicine and exercise, says nothing about the benefits of activities in the arts. A handbook on everyday cognition among the aged (Poon, Rubin, & Wilson, 1989), which contrasts real-world and laboratory research, briefly mentions Lehman's (1953) work on declines in creativity, considering losses "slight" (p. 329), but says little else about late-life creativity. Another chapter defines the reader as "a prose processor"—but nothing about literary prose. Studies of experts, whose extensive professional experience offsets age-related losses in learning and memory, excludes artists (Charness, 1989; Craik & Salthouse, 1992). An exception is a collection of readings (Adams-Price, 1998) with several chapters on aging and the visual arts, including architecture, as well as on literature and aging writers, particularly by women.

The insularity of aging studies to the arts and humanities has drawn the ire of Achenbaum (1989). "[T]he various norms and values of aging in history [and in the humanities] have been, in the main, ignored." The result, Achenbaum complained, is gerontologists "operate at cross-purposes [in designing their] research agendas [and] go about [their] business in destructive ways" (pp. xxiii and xv).

The isolation between disciplines stems from specialized (narrow) training, esoteric language (jargon), and divergent (competitive) modes of inquiry. Science puts its faith in the laboratory, adequate samples, and quantification. The arts and humanities revel in real life phenomena in all their unruliness, and celebrate free self-expression and idiosyncratic personal involvement. The commonalties between disciplines are submerged in the clash of incompatible modes of discourse.

Take the discussion in Rosand (1990) of Titian, a giant of art, as typical of the humanities' approach to late-life creativity. The spotlight falls on one exceptional old artist and one or a few of his extraordinary works, buttressed by a handful of experts' speculations about the artist's influence on others and their influence on him, and to a lesser degree, the social setting in which he worked. Richly textured personal opinions are framed by a single scholar within a highly individual analysis. The case study approach is similarly applied to other singular artists, writers, composers, and other exemplary artists and one or a few extraordinary examples of their work, depending on the scholar's particular thesis, in order to reach a conclusion distinguished from other scholars. The outcomes are interesting, sometimes fascinating, perhaps brilliant, and occasionally revolutionary.

But from a scientist's point of view, scholarly conclusions based on a single individual and a handful of works are seriously limited. Discussions of a particular old artist or a singular late work are dismissed by scientists as "merely opinions," "personal conjectures," and "unfounded speculations." No matter how insightful an in-depth study of a single case may be, it is unable to arrive at generalizations about the old age of artists as a group, late works as a whole, or late-life creativity in general. To achieve these broader goals, scientists insist that a large sample of old artists and a representative array of late works be examined. Scientists also require that the steps that gave rise to one scholar's conclusion be made explicit, open to public scrutiny, and subject to falsification, that is, capable of being proven wrong. In addition, scholars are criticized for their neglect of, indifference to, and hostility towards disconfirming evidence, for failing to move beyond personal intuition and individual hunches, and for eschewing numbers. When humanists depend on unreliable biographical accounts, uncertain historical sources, biased personal accounts, and charming but untruthful anecdotes (gossip), scientists accuse them of being "tenderhearted" (or worse). Conclusions based on "sloppy" methods, scientists claim, are inevitably flawed. Thus, the unbridled optimism of humanities about the creative potential of aging artists, untempered by any awareness of the facts about aging's decline, makes them appear polyannish, or so many scientists aver.

Scientists hurl similar accusations against clinical and psychoanalytic forays into late-life art, aging artists, and late-life creativity, much of which is favorably received in the humanities (Berlind, 1994; Gedo, 1984; Pollock, 1984; Pruyser, 1987). A therapeutic approach to the arts, critics charge, like the humanities, is "muddleheaded." A qualitative ("speculative") and selective ("loose") case study of a single patient's ("fragmentary and self-serving") reports of personal experiences ("biases") are swollen into ("unbridled and unwarranted over") generalizations, scientists' complain. Worse, still, one clinician's alleged insights on aging, elderly artists, old works, and late-life creativity, like scholars', rarely agree with colleagues, and there is no way to resolve these differences—except through exhortation, authority, persuasion, and similarly temporary means. No matter how keen the clinician, or astute the scholar, the failure to arrive at an acceptable level of consensus about late-life creativity means that it is difficult if not impossible to build upon previous efforts and move forward on some firm foundation.

Psychology's "scientist–practitioner split" illustrates the fracture between science and the humanities (Garfield, 1998). As a "hard" science, psychology follows biology and physics in content, method, and spirit. As a clinical enterprise, psychology is like the "softer" humanities: resistant to quantification and wary of strict scientific regimens.

Humanists have a rebuttal to these charges: To include many artists and works, as scientists insist, loses sight of and diminishes the uniqueness of a particular artist and blurs the peculiarities of a work (not to speak of overwhelming a scholar's resources). To concentrate on

Titian and one masterpiece, a humanist argues, preserves his individuality, the unusual facets of his painting, and the unique qualities of his creativity, all of which would disappear under the broad sweep of science. Art historians are therefore undisturbed by inconsistencies between Titian's late-life artistic creativity and that of another, 5, 10, or a hundred aging artists; one mold cannot fit all. Scholars are not discomforted by the many exceptions to the rule, or indeed, the absence of a rule. Humanists reciprocate scientists' skepticism by disdaining empirical inquiries into late-life artistic creativity. They look askance at tallies of dozens if not hundreds, nay, thousands, of creative old artists, blink a scornful eye at combining numbers for painters, musicians, writers, dancers, film makers, and other kinds of creative people in order to arrive at the abstraction "artists," and groan when comparisons are made to similarly aggregated "scholars" and "scientists." Humanists are leery, too about measuring the subtleties of old age art, the special qualities of old artists, and the complexity of elderly audience's responses to the arts, assuming these nuances could be sufficiently defined to be tallied. To treat numbers for all kinds of art as equivalent with one another and with nonart phenomena is repugnant to scholars. To consider the arts or artists homogeneously, humanists fiercely contend, loses sight of the exceptionalities of particular artists and the originality of specific works: What makes *this* older work distinctive? What kind of artist produced *this* extraordinary late work? What is an elderly viewers' *particular* experience to *this* work of art? Not: How many old works are alike, how many aging artists are similar to one another, and how identical are the reactions of the elderly to the arts? Thus, the indifference of scholars to scientific studies of late-life creativity matches scientists' disinterest in scholarly treatises on aging creative artists.

Scholars are therefore untroubled by criticisms of ignoring the scientific facts about aging and the data for older creative artists—to the extent they care or are even aware of scientific preoccupations with averaging. Scientific research on old age, despite its vaunted precision and rigor, say its detractors, is irrelevant, trivial, distorted, and worst of all, uninteresting. Scientists' blindness to the multiple possibilities of late-life creativity, critics argue, reflects their generally negative preoccupation with aging's losses. Consequently, humanists are not surprised when scientists expect, emphasize, and confirm the dire consequences of growing old. Hence, the humanities can, in good conscience, ignore attacks from scientific quarters.

Fortunately, the intransigence between disciplines can be overcome by an interdisciplinary approach to late-life artistic creativity

(Lindauer, 1998b; see also Chapter 17). The strengths of the arts and humanities—an emphasis on particular individuals and specific works—can be coupled with the muscularity of science's relentless search for the general. Thus, scientific generalizations derived from statistically massed data about old artists and late works as a group can be enlivened by narrative accounts of exemplary older artists and apt examples of their work. Science's admirable but narrow allegiance to the facts and nothing but the facts, and its inherently conservative bias against speculation and personal opinion, can be liberalized by incorporating the intuition of artists and the contextualizing of the humanities. Dramatic artistic examples and insightful humanistic interpretations temper brutish facts; quantitative generalizations come alive when illustrated by exceptional examples; and personal insights gain credibility when grounded in facts. In short, the contrasting virtues of opposing disciplines can balance their limitations and excesses.

Interdisciplinarity is difficult. Different methodologies pursuing divergent goals clash when defining problems and weighing facts. Convergencies and parallels between contrasting approaches are clouded when experts are too narrowly trained to recognize them. Mutually useful exchanges are easily disparaged when disciplinary differences are exaggerated and distorted.

However, obstacles to interdisciplinarity can be overcome in an open forum that encourages vigorous debates, welcomes contrary examples, weighs conflicting evidence, and respects counterarguments (Interrelationships..., 1980). With a democracy of ideas, discussions are invigorated, exceptions come to the fore, investigations are initiated, criticism is sharpened, lines of thought are multiplied, facts are questioned, and conclusions are reinterpreted. The results: better research, data, arguments, and analyses. Scholarly musings are therefore strengthened by scientific generalizations and enriched by exemplary illustrations—and vice versa. When divergent lines of inquiry converge they revitalize one another.

The case for interdisciplinarity, outlined above, has been endlessly promoted for centuries by many social critics, historians, and intellectuals (Klein, 1990). But little progress has been made. The concrete benefits of interdisciplinarity are rarely demonstrated, and presumed advantages remain more of a promise than a reality. Exhortations for more and better interdisciplinary study are insufficient unless followed by tangible outcomes that actually enlighten discussions in art history, extend research, facilitate the public's understanding, and lead to concrete programs.

A simple example of interdisciplinarity: Collect the scholarly analyses of old artists by different art historians, and pull together individual reports by 70- to 80-year-old artists on their creativity. Treat this "verbal material" like any other kind of data, that is, translate it into fruitful scientific questions that can be quantitatively answered: Do artists demonstrate different kinds of late-life creativity? What are they and how many are there of each kind? Which age-related infirmities matter and which don't in doing art? Do some debilities have a greater impact on late works than others? Quantitative answers to qualitative questions prompted by art professionals and fueled by artists' comments suggest further inquiries: Are expressions of late-life creativity in paintings the same as in architecture and the other visual arts? What forms of late-life creativity are expressed in poetry compared to the other verbal arts? In other words, how much overlap is there between expressions of late-life creativity in the various forms of art? And how does all this information-gathering bear on aging, creativity, and development in general among nonartists? On art history? On art programs for the elderly...? This book provides some answers.

The arts and humanities can contribute to the scientific study of aging by offering compelling arguments and clarifying illustrations that corroborate, supplement, and question the facts about late-life creativity, opening the subject to further criticism and revision. Artistic examples, artists' introspections, and audience's responses to the arts are grist for scientific exploration that lead to undiscovered facts about aging and creativity. Scientific outcomes, in turn, modify the personal observations, selected examples, and informed speculations that prompted these investigations in the first place.

The integration envisaged here requires that the hallmarks of science—systematic sampling, precise quantification, and statistical techniques—be applied to large samples of individuals and works *and* to examples of exemplary artists and their singular masterpieces, tempered by the expert testimony of art historians and other scholars, along with testimonies from artists (Chapters 6 and 7). Similarly, scholarly discussions of the creativity of a particular artist or work have to be compared to group trends for artists as a group (Chapter 8). Reports by long-lived contemporary painters on their late-life creativity are amplified when tied to factually based generalizations about aging artists as a whole (Chapters 9 and 10). Scholarly assertions about the special qualities of some noteworthy aspects of late-life art have to be tested by empirical means (Part IV). Science, the arts, and the humanities have their peculiarities, but they can work together to understand late-life artistic creativity.

There are reasons to be optimistic. Barriers between the arts, science, and humanities were not always as sharply drawn as they are today. Mathematicians were philosophers (Pythagoras), scientists were authors (C. P. Snow), artists and writers were scientists (Da Vinci, Goethe), and authors and poets practiced medicine (Somerset Maugham). A psychoanalyst, Freud, was awarded a literary prize. The founders of empiricism (Francis Bacon, James Mill) were literary essayists.

A sturdy bridge already exists between the arts, science, and the humanities. Experimental aesthetics is a scientific branch of psychology that began around the turn-of-the-century as a .highly controlled laboratory and quantitative approach to "beauty" by taking a molecular approach to the aesthetic qualities of isolated shapes, colors, and sounds (Arnheim, 1986b). Attesting to experimental aesthetics' credentials, its founder, Fechner, was one of the "fathers" of modern scientific psychology. He was a key figure in the development of psychophysics, a rigorous model for tying psychological experiences to their physical (and ultimately neurological) underpinnings. Psychophysics underlies the study of such preeminently scientific topics as sensation, sensory thresholds, and perception, and is the basis for attitude scales, personality and intelligence tests, and measurement in general wherever internal phenomena can only be measured through what we say or do, rather than directly.

Fechner therefore played a crucial role in the development of psychology as a "hard" as well as a "soft" science. Scientific studies of the arts today, or psychological aesthetics, investigates the arts, or more generally, preferences, likes, and dislikes. Following the rigorous spirit of Fechner, psychological aesthetics has contributed to the fields of cognition, motivation, and developmental psychology (Berlyne, 1974; Martindale, 1975).

Interdisciplinary barriers are formidable but not insurmountable. Reasonable compromises are possible between the artificiality of the lab, the mystery of art, the intuitions of artists, and the erudition of scholars. Rigor is not rigidity, reasonableness is not retreat, compromise need not defy commonsense. A rapprochement between disciplines is possible. The strengths of science—empiricism and quantification—and of the arts and humanities—in-depth narratives and the championing of the individual—can be joined.

In the spirit of opening up an interdisciplinary dialogue that is rooted in specific studies of late-life artistic creativity, original research is presented in this book with more information about methodology, statistics, and data than is customary in a general work. (In the case of unpublished studies, even more details are provided because they are not as accessible as published accounts.) In this informed way general readers, humanists, scholars, and artists, as well as scientists, have a concrete basis for evaluating the arguments, interpretations, and conclusions of the interdisciplinary approach offered here.

Part II examines competing arguments and evidence for late-life creativity, including views from both science and the humanities. Specifically, Chapter 3 presents the once exclusive and still prevalent decline model, in which creativity is said to decrease with increasing age. Alternative scientific evidence is reviewed and evaluated in Chapter 4. The multiple aspects of late-life creativity, drawn largely from the humanities, and in particular, art history, are discussed in Chapter 5. Parts III–V present a series of original studies on the late-life creativity of aging artists and the abilities of elderly arts audiences. The studies are set against reviews of the pertinent literature, drawing upon scholarship in the humanities, and include research participants who are art historians and artists. The book concludes (Chapter 17) with a discussion of the contribution of late-life artistic creativity to aging and creativity in general, to a positive view of the later years, and to the promotion of interdisciplinarity.

PART II

Competing Views of Late-Life Creativity

We grow neither better nor worse as we get old, but more like ourselves.

Mary Lambert Becker, 1993

CHAPTER 3

The Youthful Rise, Early Fall, and Short Span of Creativity

The Decline Model

Our nature here is not unlike our wine:
Some sorts when old continue brisk and fine.
Sir John Denham, *Of Old Age*

Few of the brightest children on an earlier era's radio program, "The Quiz Kids," are prominent today; most child prodigies fade into obscurity as they grow older. For youngsters who enter music competitions, for example, "Victory [in youth] is not a guaranteed ticket into the classical music pantheon. Politics, personal health, changing musical fashions, and the mysterious vagaries of personality, not to mention sheer luck, can be as strong determining factors as raw talent in the trajectory of a career" (Holden, 1999, p. B14). True, a few child prodigies achieved renown: Mozart, Herbert Spencer, Norbert Wiener. But in general, later fame after a promising start is unlikely. "There is no correlation between status as a prodigy and overall productivity." Thus, half of the Science Talent Search winners "do not even remain in science through college" (both quotes are from Root-Bernstein, 1999, pp. 461–462). One of the most well-known instances of youthful "burn-out" is the writer Henry Roth (1998). At age 28, in 1934, he wrote a widely acclaimed first novel, "Call It Sleep," called "brilliant" and "one of the most genuinely distinguished novels written by a 20th-century American" by critics. Yet the novel was followed by "six decades of literary silence" until Roth wrote the first of a six-volume autobiography, which was published in 1994, a year before his death at age 89. Joseph Heller wrote *Catch 22* in 1961, followed by 20 years of writer's block. Margaret Mitchell never wrote another novel after *Gone with the Wind*. Readers will probably recall classmates who accumulated awards, honors, scholarships, prizes, and other achievements but whose careers were undistinguished as adults.

Unrealized adult fulfillment illustrates a prevalent view of creativity: it occurs early in life, reaches a peak in youth, declines shortly thereafter, often precipitously, and is gone before old age. An early leader of psychology, Woodworth, claimed that "Few great inventions, artistic or practical, have emanated from really old persons, and comparatively few even from the middle-aged" (quoted in Nelson, 1928, p. 303). The best works of Shakespeare (*Hamlet*, *King Lear*, *Othello*, and *Macbeth*), Tolstoy (*War and Peace*), Michelangelo (the Sistine Chapel), and Beethoven (the *Fifth Symphony*), in the opinion of Simonton (1990a), were completed in the artist's late-30s, and even earlier, at 23, in the case of Rossini's "The Barber of Seville." Early accomplishments also describe musicians (Haydn and Pablo Cassalls), dancers (Balanchine and Graham), and scientists (Einstein wrote his famous paper on relativity at age 26).

Creativity's youthful and abbreviated span is supported by a great deal of evidence, much of it by Herbert Lehman in his seminal book, *Age and Achievement* (1953; subsequent references, unless otherwise noted, are to this book; see also Lehman, 1962a, 1962b, 1966). His analyses of the achievements of historical creative people in a variety of fields spawned the scientific study of creativity, promoted a quantitative model of research that is influential today, and gave the decline position a strong hold on contemporary views of creativity. Lehman's exhaustive findings and detailed interpretations set the tone for generations of studies showing creativity to be the province of youth, not the old. To appreciate his influence and understand the attacks and counterattacks his work generated, it is important to understand Lehman's conception of creativity and his method of study. Consequently, his work is reviewed and evaluated in this and subsequent chapters.

LEHMAN: CREATIVITY AS PRODUCTIVITY

Lehman's case rests on a rather simple procedure: count the achievements of creative figures. Tangible creative accomplishments at different ages revealed that creativity's highest point was reached between the years of 30 and 40, with a few exceptions: lyric poets were most creative in their late 20s to early 30s, and novelists peaked in their late 40s to early 50s.

The details are more complex. Lehman identified creative people in different fields by consulting authoritative sources, such as entries in encyclopedias and standard references. How many works of art were painted by Michelangelo and Rembrandt at age 20, 30, and so on until their deaths? The frequency of works-by-age, tallied for each artist, were combined so that the creativity of painters as a group could be tracked over their lifetimes. Similar profiles were generated for writers and composers, and when combined with painters, Lehman generated a profile for "artists." These were further differentiated, for example, by dividing creative writers into novelists, poets, playwrights, and literary essayists; or by dividing novelists into nationalities: American, German, and so on; or by time periods: writers of the 18th, 19th, and 20th centuries. Overall, for nearly all professions examined, Lehman found the early years to be most creative.

Consider the creative years of visual artists. Lehman examined 60 collections of "master paintings" done between 1890 and 1939 in "all of the available [books by] art historians in the Ohio University library" (where Lehman taught; p. 70). He then tallied "the frequency with which a given artist's oil paintings are listed [thereby] providing a rough indication of the greatness [creativity] of the artist." An artist's most creative work was determined by finding "The one painting... cited most frequently... for it is probably that particular artist's one best painting." The age of an artist's best (most frequently cited) work was the age at which creativity was judged highest. This procedure was followed for every historically recognized artist who worked in oil (and etchings), for sculptors, composers of grand opera, writers of historical novels and hymnals, and German writers of literature. Combining the results for all individuals in different areas of the arts established an overall pattern for creative artists in general.

The strategy of pooling frequency counts of the output of outstanding groups of individuals was applied to over a dozen professions: scientists in several areas, inventors, statesmen, and even sports figures (the ages at which prizes, championships, and first-place awards were won). The same tactic was applied to chess masters, notable figures in medicine and public hygiene, education, political leadership, and German, French, and English philosophers. When these tallies were combined, a trend emerged for creative people in general, irrespective of their particular field. Plots of accomplishments-by-ages were also obtained for less well-known individuals, the second-rankers, who were less frequently cited in authoritative sources than the more eminent. Writers mentioned only once in 10 authoritative sources, for example, were compared to first-rate figures like Tolstoy and Dickens who were included in all 10 references.

Lehman acknowledged the limitations of his procedure. It was "not an attempt to pass absolute judgment [but] merely an effort to obtain the best consensus possible...at the present time." Nevertheless, he was convinced that youthful decline was universal. "There is an optimum chronological age level for superlatively great success [and this occurs] during a relatively narrow age-range." Further, Lehman was certain that "the more noteworthy the performance, the more rapidly does the resultant age-curve descend after it has reached its peak" (all preceding quotes are from p. 75).

CRITICISMS AND REJOINDERS

Lehman admitted to exceptions: Newton's *Principia* was written at age 45, Kant's *Critique of Pure Reason* at 57, the same age at which Copernicus completed his *Revolutions of the Heavenly Spheres* (Simonton, 1984c; see also the examples of late-life achievements in Chapter 1). In addition, Lehman did not deny that some individuals who produced very little were nevertheless creative. The geneticist Mendel, wrote only seven papers; James Watson of DNA fame published only a couple of dozen papers before he became famous; Darwin and the artist Duchamp had low profiles, too (Simonton, 1990a, p. 326). Thus, not all creative people were productive—and not all productive people were creative. "Many prolific individuals are almost unknown" (Root-Bernstein, 1999, p. 460). Prolific writers can be hacks. The novelist Georges Simeon produced more than a 1,000 novels but only a few achieved critical acclaim; Melchoir Moltor (1696–1765) composed 170 symphonies (Franz Joseph Hayden composed 108; Simonton, 1984c). Complicating matters, some great intellectual achievements become a part of common knowledge and they are no longer attributed to their originators, and hence not "counted." In short, tallies of productivity will no doubt miss some creative people and include a few unworthy ones as well.

The emphasis on productivity-as-creativity raises other important issues. Is quantity equivalent to quality? Does an external index, like output, reveal underlying processes or traits? When productivity is called creativity, how much more is known about creativity: What then determines productivity? Other perplexities: The reputations of individuals change with time, some rising, others falling, thereby affecting what is counted for whom and when (and what did the original counts signify?). Information about the accomplishments of historical figures is often sparse, frequently questionable, and sometimes contradictory.

To Lehman's credit, he recognized these and other problems and attempted to answer them meticulously, point by point, often spiritedly, and occasionally with some pique. Objections by critics "are largely a result of...misunderstanding my book [and are therefore] invalid" (p. 334). Critics who called attention to exceptions, rejected his averages, and insisted on analyzing individual achievements in detail were, "utterly naive" (p. 144). In an unbecomingly heated, personal, and vituperative manner, Lehman raised doubts about whether one of his fiercest critics had even read his book or had read it selectively. "Can [he] really be serious?" (p. 336). In a petulant exchange over a particularly critical set of data (table 58, p. 317), Lehman argued that his critic "overinterprets" them, then admits, "perhaps I may be partly to blame," and offers a lengthy excuse about not having sufficient funds to publish more information. (All quotes are from Lehman, 1956, p. 335.) (Nevertheless, in several subsequent publications by Lehman, as far as I can tell, the missing information was never provided.)

Lehman acknowledged exceptions, but they did not trouble him. Great figures erroneously omitted, he argued, can be added and unworthy ones unjustifiably included can be deleted when more knowledge becomes available. Omissions and errors were therefore only temporary. In any event, Lehman maintained that exceptions were too few to markedly affect general trends, which were based on hundreds of acknowledged creative individuals from a large number of different fields in dozens of countries, several centuries, and representing thousands of recognized works.

To the charge that his conclusions were contaminated by the inclusion of individuals who died at different ages, making it impossible for individuals who died young to make a creative contribution at older ages, he snapped, a "groundless allegation." He was correct. One of Lehman's many tables takes longevity into account (p. 334), and indicated that age-of-death did not affect the curve of creative production. "Even if all [his emphasis] of the creative workers had been long-lived, the relative creative production rate of the oldsters would not have been as great as that of the younger men." For example, artists who lived to at least 70 differed little from short-lived artists in the pattern of their creativity (figure 55, p. 79 and figures 53–54, p. 78; a different analyses of these data for artists, and another interpretation, are presented in Chapter 4).

Lehman's point about the irrelevance of longevity was partially corroborated by Dennis (1955, 1966), who counted the productivity of only long-lived individuals in the arts, sciences, and scholarship. Decline did occur early, in the 40s, for noteworthy individuals who lived until their eighth decade. Note, though, that this decade is some-what later than the 30s, the period favored by Lehman. Dennis also observed that artists' productivity fell earlier—and more sharply—than other professions (a point I will return to in Chapter 4).

Lehman also defended his reliance on group trends rather than individual cases. "Human death [is an individual matter]," he con-tended, "but this does not prevent actuarians from assembling mortal-ity tables." Lehman conceded, though, that group averages leave "room for many exceptions." To demonstrate his fairness, Lehman included a chapter on "precocious youths who produced remarkable creative works before they were 22 years old" and another chapter on "distin-guished individuals who announced their most famous achievements in late old age." This finding, Lehman quipped, "should provide encouragement for oldsters" (p. 144; the earlier and following quotes are also from this page in his 1953 book).

Lehman was open to the *possibility* (his emphasis) of uncontrolled or extraneous but unavoidable influences that might have affected his findings. The birthdates of some individuals who lived long ago, he admitted, may be unreliable or approximate. Scientists' attitudes to publishing have changed, too, and could skew publication dates and rates. Scientists also censored their work in earlier times due to fears of being charged with heresy. Harvey and Darwin, for example, were reluctant to publish controversial material. Some inventors and scien-tists kept their work secret as long as possible because of the absence or weakness of copyright and patent laws. Some paintings were prob-ably done by students of the masters who signed their names and received sole credit. Lehman therefore accepted the likelihood that some of his data may be distorted, and that unknown and indeter-minable factors could push frequency counts either up or down.

But Lehman steadfastly held that these problems were minor and did not seriously distort his general findings. Lehman grudgingly con-ceded that new data might modify his interpretations slightly but would not change them. "Adverse criticisms are unsustainable," he maintained. For "a comprehensive picture of the over-all facts," he advised readers, take into account "the consistency of the data across many samples, and not just one." In short, Lehman's critics are simply "wrong" (quotes here and below are from pp. 147, 335, and 337).

To Lehman's credit, he did not equate age itself as the cause of decline. Aging, he argued reasonably, brings about losses in physical vigor and sensory capacities, increased illness, decreased mental abilities, greater rigidity, and a shift to more practical matters. These age-related events, not age per se, affected late-life creativity. "The age factor of itself could hardly be regarded as *causing* anything...the decrements in the creative production rates occur *with* but certainly not *because* of the passage of time" (emphases are his).

WHAT ABOUT PRODUCTIVITY?

The most serious complaint against Lehman is his exclusive reliance on productivity counts as indicators of creativity. He defended this choice by arguing that some product, outcome, or tangible marker is the only way to demonstrate an individual's creativity. Productivity measures have their limitations, Lehman conceded, but they were considerably fewer than allusions to "the creative mind," "the creative spirit," "the creative personality," or "the creative type" (pp. vii, 70–72).

> The product, broadly defined [to include expressed ideas as well as a poem, dance, essay, cake, computer program, or machine tool] seems synonymous with creativity [for it is] the very thing that allows us to determine whether a person or process is creative. (O'Quinn and Besemer, 1999, p. 413)

Others have also made a strong case for requiring a material product to identify creativity (Albert, 1975).

A helpful distinction was made by Root-Bernstein (1999, see especially pp. 459–460) between total productivity, which includes everything (the complete works of an artist) and effective productivity (works that are significant and exemplary, such as the art in museums). Effective productivity, Root-Bernstein argued, was emphasized by Lehman. Further, the two kinds of productivity were manifested at different ages. The more significant works, according to Root-Bernstein, are produced earlier in life, and these "revolutionary," original, and ground-breaking efforts are expanded upon later in life, resulting in total productivity. Defining creativity in terms of a product also has the advantage of focusing on real-world accomplishments, which are more direct and compelling than, say, test scores of creativity. Actual achievements over a lifetime are not like the fleeting accomplishments of school children and adolescents.

Reassuringly, productivity is not as crude an index of creativity as it might appear. Output correlates with self-evident measures of creativity, such as the judgments of experts, the number of awards, prizes, and other signs of recognition, creativity test scores, and a wide variety of measures of achievement (Simonton, 1984c, 1986, 1988b). Productivity counts are also handy pegs on which to hang other indicators related to creativity, both qualitative (reputation) and quantitative (the number of citations to a work by authorities). Peaks, plateaus, and declines in creative productivity can therefore be linked to social, cultural, economic, and historical events, as well as personal records, like birth order, family size, number of children, presence or absence of parents, and a host of other variables, even the seasons of the year. Correlations also suggest possible contributors to creativity, ranging from the influence of parents and peers to economic and historical factors.

DECLINING CREATIVITY TEST SCORES WITH AGE

Lehman's studies of the decline in creativity-as-productivity, based on historical figures, have been confirmed by creativity tests of living and non-eminent persons (Alpaugh & Birren, 1975, 1977; Alpaugh, Parham, Cole, & Birren, 1983; Bromley, 1956; Diamond, 1984; Horner, Rushton, & Vernon, 1986; cf. Cole, 1979; Smith & Kragh, 1975). A good example is Ruth and Birren's (1985) study of the creativity test scores of three age groups: a youthful 25–35, a middle-aged 45–55, and an old sample of 65–75. Participants completed four tests of creativity with both verbal and nonverbal (pictorial) materials representing diverse theoretical viewpoints: Guilford ("Uses Test"), Torrance ("Just Suppose" measure), and Wallach and Kogan ("Visual Patterns"). The tests measured three aspects of creativity: "fluency," the total number of responses; "flexibility," the different categories into which responses fell; and originality, the number of unique responses. The atmosphere in which the tests were given was also varied: either informal (relaxed, supportive, and cheerful) or formal, the traditional testing situation, where instructions were printed and read by participants rather than, as in the informal setting, told to the participants. Some groups also received the creativity tests after a test for intelligence, a condition intended to elicit anxiety.

Creativity differed by age on all measures, "to the disadvantage of the old"; the youngest group had the highest scores. The age-related losses were due, Ruth and Birren speculated, to older subjects' "reduced speed of information processing, less ability in handling complexity, and a decreased willingness to risk making original solutions," along with educational, occupational, and social handicaps. Surprisingly, though, the difference between middle-aged and the oldest group was "rather small"; differences were greatest between the young and middle-aged groups, not between the young and the old (all quotes are from pp. 99 and 104). (The informal testing situation was more beneficial, although not to the advantage of any age groups. Similarly, gender differences were not age related.)

Another exemplary example is the study by McCrae, Arenberg, and Costa (1987). They tested individuals aged 17 to 101, relying on both a cross-sectional and longitudinal research design; the latter was conducted between the years 1959 and 1972. The investigators used six tests of divergent thinking, tapping a kind of mental agility typical of creative behavior where more than one correct answer to a problem or question is possible. (In convergent thinking, the more traditional response, especially in school settings, only one answer is correct.) Divergent thinking declined with age on all the tests, and like Lehman's results for productivity, was sharpest after age 30.

SUMMING UP

Lehman's influence on research and thinking about the course of creativity cannot be overstated. Opponents of the decline position begin by re-examining and re-evaluating Lehman's contributions (Chapters 4 and 6). The declining arc of creativity with age is largely accepted by the major figures in the field, including Simonton, one of the most prominent. "The probability of conceiving notable creative products tends to diminish as the creative passes the peak productive age of around 40" (Simonton, 1984b, p. 93). "Creativity seems to peak in early or middle adulthood" (Simonton, 1990a, p. 321). There is "a decline in creative activity with increasing age" (Simonton, 1994, p. 182). Even critics of Lehman grudgingly concede that their objections have not been "sufficient to change the basic conclusion [about youthful decline]" (Mumford and Gustatson, 1988, p. 29).

Lehman's achievements cannot therefore be ignored, especially since they have been expanded by proponents of productivity counts and adherents of the decline position. Simonton again: "Research has tended to endorse [Lehman's] chief conclusions [and] nonmonotonic curves appear again and again throughout the literature with a robustness seldom seen in the behavioral sciences" (Simonton, 1990b). In a rare but muted demurral, Simonton noted that Lehman's main conclusion has to be somewhat qualified because chronological age is not the same as career age. Nonetheless, decline is usually the case whether expressed in terms of career or chronological age. Lehman's work has been successfully extended and not "flatly contradicted" (above quotes are from pp. 88 and 103). Despite some qualifications, Simonton's resounding last word is that the "age decrement in the last years is real," and confidentially declared, "Almost without exception, investigations into longitudinal changes in performance on [psychometric indicators like tests] show a decline with age...and in no instance can creative capacity be said to increase significantly with aging in the last half of life" (Simonton, 1990a, p. 326). Based on his continuing and impressive series of mathematically elegant and statistically sophisticated studies, Simonton builds upon and affirms Lehman's work (Simonton, 1997).

Whatever its shortcomings, Lehman's work promoted an empirical, objective, and reliable way of studying the course of creativity, and for disclosing systematically, for the first time, the pattern of creative development. His approach led to a great deal of information about creative output in a number of different fields over a wide swath of history. Lehman's contribution deserves to be called creative. Ironically, criteria other than productivity would have to be used since Lehman was not a prolific writer!

Lehman's approach to creativity has withstood many challenges. His seminal finding—that creativity peaks in youth and fades well before old age—has repeatedly been confirmed for historical and contemporary figures, and for occupations he did not initially examine, mainly areas of science, especially physics and chemistry, engineering, as well as other fields, including the arts (Simonton, 1984a, 1984c, 1990a). Tests of creativity and other measures also confirm a pattern of declines with age. Thus, publication rates and citation counts for contemporary scientists and mathematicians display a pattern of youthful decline that matches the historical record uncovered by Lehman.

Decline is therefore the generally accepted view of creativity (Kastenbaum, 1991b). Creativity rises "between the thirties and fifties for most individuals and most disciplines [and] then declines with age" (Root-Bernstein, 1999, p. 459). Rowe and Kahn's (1998) seminal book on successful aging accepts decline as a fact (see Chapter 11 on productivity), although they temper its severity by pointing out, as Lehman and others have, that the age of decline depends on the field (mathematicians peak early and philosophers late), and emphasized "large" individual variations (p. 129), which Lehman called exceptions.

REASONS FOR DECLINE

Creativity's decline is therefore generally accepted and the issue has shifted to its explanation: What happens between youth and old age that causes this decline in creativity? Numerous reasons have been given. Lehman himself suggested at least 16: Losses may be the result of changed work conditions with increasing age, as administrative, familial, community, and professional duties increased and interfered with the creative process itself, its expression, or both. The need for achievement and success, which fuels creativity, Lehman continued, may also decline by mid-life after a sufficient amount of material goods have been acquired. Striving for creative achievement may therefore lose its driving force as a reinforcer/reward for creativity. Similarly, Lehman adds, societal pressures for status and prestige decline in importance with increasing age, thereby weakening or eliminating the underlying motive for creativity. Decreased motivation and ambition, furthermore, are accelerated by failing health, physical impediments, and the sensory handicaps of aging.

About a dozen more reasons for creativity's decline were added by Abra (1989, 1995): Decrements occur in thinking, mental flexibility, and memory; persistence weakens; practical needs take precedence over fresh impulses; self-confidence decreases; awareness of mortality increases; and other concerns receive a higher priority than creative work. Detrimental to late-life creativity, too, is the loneliness and insecurity that comes with age, and a shift in gender roles, which according to Abra, has men becoming more feminine and hence less creative.

Still other explanations of declining creativity apply to specific professions, especially aging artists. According to Kastenbaum (1989, see especially p. 103) artists are more aware, sensitive, and responsive to aging and death than nonartists. The result, Kastenbaum argued, was a depressed state that negatively impacted late-life creativity because it fed an urgency to produce that undermined creativity. Also frequently mentioned as especially detrimental to aging artists are sensory and physical losses (Linksz, 1980; Trevor-Roper, 1970). Mary Cassatt's blindness was obviously fatal; and motoric impairments obviously impair musicians, actors, and dancers. Mentioned frequently among artists, too, are depression, suicide, and other forms of pathology that are exacerbated with age (Jamison, 1994; Lindauer, 1994; Pollock, 1984; Raymond, 1989; Simonton, 1986). Personal crises late in life also detract from artistic creativity. Aging artists, in overcoming religious doubts and spiritual uncertainties, either repeat themselves or attempt to become more innovative, tactics that work against creativity (Gedo, 1984).

A provocative explanation, more than 100-years old, was offered by Beard in 1874 (revived by Simonton, 1977). Beard examined the lives of 400–500 distinguished men and women from at least seven different fields and discerned two competing trends with age: a waning of youthful enthusiasm, and in opposition, an increase in experience. Aging provided the experience that allowed originality to be expressed in an intelligible fashion, but also brought about a decline in enthusiasm and a preference for the routine, both of which reduced the benefits of increased experience. With enthusiasm peaking early in life and experience increasing with age, creativity was maximized when the two opposing forces were balanced. An optimum intersection occurs around middle-age, as Lehman would expect, although Beard placed creativity's peak slightly later. "Seventy per cent. of the work of the world is done before 45, and eighty per cent. before 50 [and] on the average, the last twenty years in the lives of original geniuses are unproductive" (p. 7).

Beard's model was critically evaluated by Mumford and Gustatson (1988). If Beard is correct about experience increasing as enthusiasm declines, why are any middle-aged men more creative than youthful individuals? Mumford and Gustatson contended that even relatively minor creative efforts, whatever one's age, require substantial enthusiasm. They were therefore unconvinced of decline's maximization at mid-life when knowledge, social connections, and professional expertise were greatest and expanding.

Despite their objections to Beard's explanation, Mumford and Gustatson accept the fact of decline but attribute it to oppositional forces, variously called homeospatial, Janusian, and cross- and multiple-thinking by the theorists Koestler (1964) and Rothenberg (1996). What these terms have in common, Mumford and Gustatson claim, are tensions between disparate influences. These are resolved by creative individuals, according to the authors, by reorganizing conflicting facts and ideas through a gestalt-like synthesis. The need for and manner of synthesizing, Mumford and Gustatson argued, differ for the old and young. In youth, there is a better accommodation between inner and outer tensions; hence, creativity peaks. But with age, and an increasing awareness of death, there is the realization that synthesis cannot be achieved; thus, a decline in creativity.

Why then do more young people not show bursts of creativity, and why are some old people creative? Mumford and Gustatson answered: When the old act like the young, by adapting and accommodating to new information, competition, and conflict, they remain creative; and when the young act like the old, satisfied with the status quo and content with preexisting knowledge, creativity either fails to emerge or ceases. How to account for the different ages at which creativity peaks in some professions? Mumford and Gustatson explained exceptions as due to the interplay between innate and environmental forces. "Curves tend to shift downward in fields highly dependent on native ability (e.g., youthful mathematicians) and upwards (for older novelists) in fields requiring substantial training and life experience" (p. 29).

None of the many explanations of creativity's decline with age, however, is satisfactory. "To date, no hypothesis has been proposed which can provide an adequate explanation [for decline]," concluded Mumford (1984, p. 229). The many contending accounts are problematic, too. Not all can be correct. Are some better than others? The proliferation of explanations may reflect attempts to solve questions left unanswered, or perhaps they address different kinds of decline—or creativity.

Whether there is one, several, or many explanations of creativity's decline, the fact is that an impressive body of evidence demonstrates that peaks occur relatively early in life, last only about 10 years, on average, and fall between the 30s and the 40s in most professions. These data are rarely questioned but taken for granted. Other possible trajectories are hardly considered for several reasons: a similarly negative pattern of losses describes aging in general; higher order cognitive abilities, like intelligence, decline with age; a generally pessimistic attitude is taken to human potential; and psychopathology exclusively focuses on illness rather than health and optimism (Chapter 1).

Age-trends for creative artists appear particularly discouraging (Cohen-Shalev, 1989b). "The general opinion among life span psychologists is that, on the whole, allowing for a few exceptions, aging is not favorable to artistic creativity.... Psychologists of various theoretical backgrounds, as well as art critics and literary scholars, seem to argue that there are not a great many creative individuals who have produced noteworthy significant contributions in their old age" (pp. 34–35).

The decline model of creativity has conceptual, factual, and methodological difficulties—just as accounts of aging do. Lehman's approach to creativity, with its emphasis on productivity, has disturbing assumptions, unresolved questions, and controversial answers. The decline position has been challenged by contrary empirical research findings, examples of individuals whose creativity continued into old age, and counterarguments by scholars in the humanities, particularly art historians who write about late-life art. The critics do not deny that creativity often declines with age or even that it predominates in human development. But they insist that other views of the course of creativity need to be considered. This is especially true for aging artists and the art of old age. These alternatives are presented in this book, but first a critical examination of the decline position.

CHAPTER 4

Does Creativity Decline with Age?

Let us cherish and love old age; for it is full
of pleasure, if one knows how to use it....
The best morsel is reserved to the last.

Seneca, *Epis.*

Late-life creativity is underrepresented in gerontology and psychology (Chapter 1), and when the topic does receive attention losses are emphasized. With aging, creativity falters (Chapter 3), echoing a general stance toward aging as a time of decline (Chapter 1).

Yet many older individuals in a variety of fields are creative (Chapter 1); decline and age are not necessarily synonymous. Further, decrements in creativity with age are not explained by age, but rather, by certain factors related to or concurrent with age. Even Lehman (1953), the major proponent of the decline position, asserted that creativity peaked for some who were older than 30 and 40. These and other problems with the decline model are critically examined in this chapter, especially the methods used to investigate it and the attitudes that underlie it. Research in opposition to this model is reviewed in subsequent chapters.

PROBLEMS WITH THE DECLINE POSITION

The multiple explanations of the decline of creativity (Chapter 3) are a sign of disarray, since good science is characterized by simplicity (parsimony). Troublesome, too, are explanations that work equally well against decline as for it. Take an increased awareness of mortality. Creativity might be stifled by a gnawing preoccupation with death. Alternately, as life nears its end, new heights of creativity could be spurred as a lasting testament, as in composers' final creative "swan song" (Simonton, 1989). A third possibility: the imminence of death could leave creativity unaffected. The noted author James Mitchener was asked at age 83, after having written 10 books in four years, whether an awareness of mortality affected his work (Keever, 1990).

Mitchener replied that it had not. When he asked himself, "Was I …apprehensive of work-ending death?…did I labor so diligently because of my age and the approach of a time when I could work no more?" his answer was, "I think not. I write at 83 for the same reasons that impelled me to write at 43" (p. 3). The reasons he gave were to communicate, to organize experience, and to tell stories that dramatized adventures readers might have had.

Another difficulty with the decline position is that it occurs in the 30s and 40s, when health, sensory capacities, and the intellect are at their prime, or even earlier at a more vigorous age for poets and mathematicians (Mumford & Gustatson, 1988). Indeed, most people are quite strong, fit, and mentally competent in their 40s and 50s, the years in which creativity supposedly has disappeared. Only in the 70s and 80s, when physical strength has markedly diminished and illnesses increased, would creativity be expected to be seriously impaired for reasons of health.

A host of personal, familial, and various other burdens work against creativity at every age. Why should they be more devastating in the early and middle years rather than later? In youth, when creativity flowers, severe pressures exist: breaking away from family, succeeding in school, choosing a profession, establishing a career, making a living, developing new relationships, becoming socially active, getting married, starting a family. Competing stresses, interferences, encumbrances, blockages, and deflections occur at every age, from the earliest years, through adolescence, young adulthood, and the middle years, and into old age. There are therefore many reasons for creativity to decline at every age, not only in old age.

When losses in strength and other abilities do occur, they do not necessarily bode ill. Creativity does not require hard work and strenuous labor, and odious and tiresome tasks can be carried out by assistants. The handicaps of aging can be ameliorated by reorganizing priorities, revising goals, developing helping relationships, and building a network of useful contacts. Slowing down because of age can lead to the discovery of new materials, efficient labor-saving devices, better techniques, and unusual styles. Revised coping strategies can serendipitously lead to new forms of creativity. The aging Matisse turned to large cut-out collages when he could no longer hold a brush. When Rembrandt grew older, his "hand travels less stylistically, conserving its energies, drawing more economically [and becoming] more deliberate and cautious …. His movements, those of an old man now, are also those of the characters he projects" (Rosand, 1987, p. 92).

A number of aging artists suffered from physical, motoric, and sensory debilities, but these did not impair their art (Bergmann, 1972;

Cowley, 1980; Linksz, 1980; Trevor-Roper, 1970). A looser brushstroke due to losses in hand coordination, for example, is not considered a detriment to artistic expression. Poussin painted successfully despite a trembling hand; and Renoir strapped a brush to his arm after being disabled by arthritis. When Goya's eyesight failed at 78 (he was also deaf), he wore several pairs of glasses, supplemented by a magnifying glass (Berman, 1983). Rubens suffered from gout and relied on assistants. "Faced with Ruben's late works without information about his life, it is unlikely that one could detect any physical disability" (White, 1977–78, p. 53).

Several of the Impressionists had severe medical and physical problems, but their paintings were "artistic triumphs over physiological disorders" (Hamilton, 1984, p. 237). Creative ability is independent of physical deterioration. "Intense artistic creativity can inhabit wasted bodies just as the body can long survive the loss of mind" (Held, 1987, p. 128). Four or five of seven major Impressionists, according to Ravin (1985), were afflicted in mid-life or later with ocular disorders that affected their perception of color, shape, and space. Monet's cataracts, for example, influenced the way he used certain tints. Other neurological deficits, Ravin added, accounted for changes in the late works of other artists. Artistic *expression* may be affected, not creative *potential*.

Monet and Degas are illuminating examples. Both lived and worked around the same time, had long and successful careers, died in their 80s—and had progressive eye diseases (Ravin & Kenyon, 1998). Monet's cataracts made it difficult to distinguish colors and forms. "His paintings of waterlilies and weeping willows done between 1918 and 1922 reveal less and less [violet and blue tones] and more yellow, reds and brown." Degas' ocular problems were magnified by depression. His "late human figures are colder, less animated, more featureless than his earlier work." Nevertheless, "despite their physical and psychological problems [Monet and Degas] created masterpieces in their late years" (quotes are from pp. 257 and 266).

In Sum. Age-dependent losses can stagnate and exhaust creativity, and for those so afflicted, decline is the result. But increasing age also leads to more mature thinking, revitalizes ideas, and opens new avenues of expression, and for those so blessed, creativity is enhanced.

PRODUCTIVITY TALLIES AND TEST SCORES AS INDICATORS OF LOSSES IN CREATIVITY

The main indicators of creativity's decline with age are reduced productivity and lowered test scores. The problems of both measures, noted earlier (Chapter 3), are examined further here.

The Problems with Productivity

Productivity is an ambiguous term because it can be measured in several ways (Root-Bernstein, 1999). The creative productivity of teachers, for example, might be tallied by counting the number of hours of teaching, courses taught, handouts developed, pupils educated, students who go to graduate school, or graduates with successful careers. Several quantitative benchmarks identify a productive businessman: Profits, worker output, market share, and stock prices. To stretch a point, creative parents might be defined by the number of children they have. Further, total productivity is not synonymous with valued productivity; and the latter is distinct from creative productivity, and these different kinds of productivity may decline at different ages. It is difficult to evaluate the quality of quantity and the equivalence between declines in productivity and losses in creativity.

The singularly powerful effect of a single creative work poses another problem for a productivity criterion. One book, Darwin's *Origins of the Species*, or Freud's *The Interpretation of Dreams*, drastically altered several fields of inquiry, if not a civilization's worldview. An individual recognized as creative may produce very little (e.g., Duchamp in art). Lehman (1966a), the chief advocate of the productivity criterion of creativity, recognized the importance of a single work. "The productive rate as found for a group of contributors and individual production rates are two entirely different things...there are many individual exceptions to the general group trend" (p. 270). (On the same page, Lehman made another startling admission: "An individual may do highly important creative work at *any* time from his late teens and into his eighties" [emphasis added]. This from an advocate of the early rise and fall of creativity!)

Productivity counts, as large as they can be, present a paradox. No matter how huge the number of greatest figures and most outstanding works, a relatively small amount of works and individuals are tallied: How many seminal works are there? Renowned persons? "Giants"? A few dozen, a few hundred, or even a few thousand is a finite number that may represent only a fraction of what could be counted. Thus, one of Lehman's (1953) major data sets, the "world's best" paintings, numbered a scant 40 and were by only 28 artists. Productivity counts therefore overlook less well-known creative individuals—a van Dyck as well as a van Gogh—who are important and influential. The output of a relatively few geniuses and masterpieces, the breakthrough's and revolutionary moments in history, the most creative in a profession, underrepresent relatively lesser but renowned creative individuals.

The patterns of creativity over time may differ for the first, second, and third ranks.

No matter how much of the creative output of many individuals are counted, the characteristics of the tallied works are ignored, the qualities that make individuals and products creative. What exactly is it about a work of art, or a scientist, that makes it or the person creative? A collective index of the creative output in a profession says nothing about the particular virtues that made the tallied works endure. Was it an artist's new way of exploring an old problem or the application of an old approach to a novel situation? Did the artist treat light or handle color in a unique way? Thus, general productivity averages cannot reveal the groundbreaking and trendsetting qualities of originality, effectiveness, specialness, and lastingness that made the works countable in the first place.

Similarly, productivity counts reveal nothing about the lives, working habits, family, friends, relationships, and social circumstances that influenced the productivity of creative individuals whose works were tallied. Averages do not illuminate the pressures of the marketplace, the compromises resisted in earning a living, the sources of support and patronage, and the working circumstances that enhanced creative output, or the changes in these conditions over time.

The Problems with Tests

Creativity test scores, like productivity counts, decrease with age (Chapter 3)—and have their share of problems. They have all the difficulties of any test, whether it be of personality, intelligence, and the like (Hocevar, 1981; Romaniuk & Romaniuk, 1981). Creativity scores obtained in youth may decline with old age not because creativity has decreased but because the test instrument was not reliable. A creativity test may not measure creativity, but rather, test-taking ability (a problem of validity). The elderly, like minority and other underprivileged groups, perform more poorly than the young because of fewer advantages (Butler, 1963, 1973; Labouvie-Vief, 1984). Young people still in school are more familiar with tests, especially the multiple-choice and short-answer kinds that are timed and have only a single correct response. Elderly test-takers are handicapped because they are less likely to have faced these tests as recently and extensively as the young, are probably less experienced with verbal and psychological tests, at least since they left school, and have probably never taken a test of creativity, which were only relatively recently developed and

widely used. Tests of creativity, moreover, are developed, standardized, and pre-tested with younger populations to whom they are
mainly addressed. Nevertheless, some older individuals score higher
than some younger ones.

Low scores may also reflect the unique history of different age
groups: Young tests-takers were born after TV and computers were
invented; the middle-aged lived through the Korean, Vietnam, and
other wars; and an old group experienced the depression and World
War II. The unique historical events associated with each age, along
with differences in education, values, and attitudes connected with particular time periods, account for differences in test scores rather than
reveal losses in creativity. Thus, reduced creativity test scores over time
may illustrate a socioeconomic rather than decline model of creative
development (the problem of construct validity). High scores on a
creativity test might not have much relevance to real-world creativity,
like producing a masterpiece of art, making a scientific discovery, or
developing a useful invention, and are therefore not as compelling as
real-world creative accomplishments (face validity). Test scores are also
uncertain predictors of future creativity. What do high-scoring students
actually accomplish when they become adults? Creativity tests,
obtained under artificial conditions, capture only a brief moment of
time in a particular setting. Low creativity test scores among the elderly
may therefore be the consequence of one or more factors, unrelated to
a decline in creativity.

A way to resolve many of the difficulties with tests and other
measures is to compare the creativity scores of the same 20-year-olds
when they are 40, 60, and 80 years old (longitudinal research) rather than,
as is more typical, compare different 20-, 40-, 60-, and 80-year-olds (cross-
sectional research). The latter decline in creativity while the former show
either constancy or gains with increasing age (see the studies by Terman
and his successors, and by Erikson and his coworkers, Chapter 1). Despite
the longitudinal methods' advantages, cross-sectional studies dominate
because of their convenience (Subotnik & Arnold, 1999). Snapshots of
different aged people at a single time are easier to obtain than an on-going
study that requires many years to complete before a meaningful pattern
emerges. Participants in longitudinal research, furthermore, die, drop
out, or lose contact with researchers as time passes. For these reasons,
cross-sectional research, with their interpretive ambiguities, and paper
and pencil tests of creativity, with their problems of reliability and
validity, are almost exclusively depended upon, and they support the
decline model of creativity. More subtle factors also maintain the
dominance of productivity counts and creativity tests, discussed next.

DECLINES IN LATE-LIFE CREATIVITY: THE IMPACT OF BEHAVIORAL MEASURES AND AGEISTIC ATTITUDES

Creativity tests and productivity counts in support of the decline position have a firm grip on studies of late-life creativity despite their methodological and conceptual shortcomings. Their dominance rests on scientists' preoccupation with behavioral measures, and less obviously, ageistic expectations about aging. These biases hinder a more balanced approach to late-life creativity and prevent a consideration of options other than decline. For example, declines in productivity tallies and test scores with age can be accepted without denying that creativity can exist as a potential, an underlying trait, or a covert process, albeit not easily countable or neatly scorable. Declines in overt measures as age increases can therefore be interpreted as declines in creative *behavior* or *performance* rather than in creative capacity. Thus, the outward signs of creativity are diminished by physical illness, sensory losses, and other handicaps of aging, but not creativity itself. Accordingly, late-life creativity can exist covertly, even though creative productivity is absent and test scores are low (Schaie, 1990). Losses in creativity are therefore only apparent when concrete manifestations (test scores, productivity counts) are the sole criterion.

Many scientists, still, reject alternative interpretations of declining productivity and lowered test scores, arguing that the *absence* of a behavioral measure cannot indicate the presence of something. This positivist-behavioral orientation is difficult to overrule, despite cogent and compelling arguments. Scientists have an unwavering faith in overt indicators, like scores and frequencies, as the best (if not only) way of studying covert phenomena like creativity (or personality or intelligence). They might argue as follows: "Critics of behavioral measures have a point and may well be right. But how else can unobserved, unobservable, and unexpressed phenomena like creativity (or motives or attitudes) be discerned in youth and old age? What is the alternative?"

The rejoinder might take the following form: "Stereotypes and prejudices about old age, collectively known as ageism, lead to preconceived notions about the relationship between aging and creativity. These expectations affect the methods which investigators use to study age and creativity (tests and tallies), the hypotheses to be tested (losses), and the interpretation of the results (decline)." If creativity in old age is not expected, the critics continue, it will not be looked for, and alternative methods other than tests and tallies and scores will not be pursued, like questionnaires and interviews, and possibilities other than losses will not be found (like gains and constancy). It is perhaps with these points in mind that Dennis (1956) faulted Lehman's productivity counts for implying early "deterioration" and for their "deleterious consequences" on views about aging (p. 331).

Preconceived notions about aging and creativity therefore prevent researchers from looking for more benign aspects of late-life creativity, including the following: Living long increases the odds of producing creative works, provides more opportunities to make up for past mistakes and failures in creative efforts, gives critics time to change their minds about an individual's apparent lack of creativity, and allows past works to be appreciated for creative qualities that were overlooked or rejected earlier.

Ageism, rather than productivity tallies and test outcomes, may therefore account for the neglect of late-life creativity and an overemphasis on its losses. "This belief system [ageism] has created a vicious cycle, generating evidence in accordance with it and ignoring evidence to the contrary," a prejudice that is particularly hard on artists: "The number of successful aging artists [is] conveniently underestimated [and] their late contributions have not been given a fair trial" (Cohen-Shalev, 1989b, quotes are from p. 35).

When ageism is combined with sexism, the result is fewer studies of old women artists (Rodeheaver, Emmons, & Powers, 1998). When ageism and sexism are combined with romanticized myths about creative people (Lindauer, 1994)—they have to suffer, they are mad, they live only for their work, and they die young—the expectation that creativity disappears with age is strengthened even more.

PROBLEMS WITH LEHMAN'S RESEARCH

Beleaguered advocates of the decline position might cry out at this point, "But what about Lehman's data?" The decline position draws much of its strength from Lehman's comprehensive tallies of the productivity of historical figures in various professions. His findings are constantly referred to in research and theory, receive a prominent place in textbooks and other standard sources, and strongly influence the study of creativity. The robustness of Lehman's findings is unusual for research that was completed over 50 years ago; data more than five years old are usually suspect. Yet the seminal work on successful aging by Rowe and Kahn (1998) accepted Lehman's findings and the decline position on creativity (see chapter 11, especially p. 129) with only minor exceptions (Chapter 3). More puzzling, though, was Rowe and Kahn's failure to recommend greater involvement in the arts as a way of achieving a successful old age, and their omission of art when discussing the ways in which leisure time might be spent (they promote volunteering). Ironically, Rowe and Kahn referred to Michelangelo and Picasso as illustrative of late-life productivity; and their chapter on productivity, with its emphasis on the value of a long life, was introduced by a quote from *Ulysses*, a poem by Tennyson.

Despite the widespread and enthusiastic acceptance of Lehman's work, there are serious questions about productivity counts, creativity test scores, and cross-sectional research designs that cast doubts on decline to the exclusion of other possibilities (Denny, 1982; Levy & Langer, 1999; see also Chapters 3 and 6). A particularly strident attack came from Dennis (1956). (For his reply, see Lehman, 1956.)

The samples Lehman used, Dennis argued, were inadequate in size and representativeness. The age intervals chosen, furthermore, were arbitrary, changing, and small. For example, Lehman's choice of a 5-year span was oversensitive to small differences. Dennis also questioned the adequacy of Lehman's historical sources: They cannot be combined in making tallies because the individuals and works in each depend on a variety of unknown criteria of creativity and productivity. Worse, they were outdated. For example, Lehman's sources for artists and their works were published in the first part of this century or earlier. Ideological, aesthetic, and other standards about art have shifted markedly since then. Consequently, the authorities Lehman consulted in his choices of great art and gifted artists were out of date.

Dennis listed other criticisms. Scientists and scholars from the past, unlike today, were not pressured as much to "publish or perish." Hence, less urgency to be productive. Compared to the present, too, there were fewer scientists, artists, and other professionals, and hence fewer products to count. Thus, a relatively small number of creative individuals from the past, Dennis complained, have a disproportional impact on productivity trends.

Dennis also faulted Lehman for relying on data that were biased in favor of the younger years (cf. Lehman, 1958). Thus, the first great painting by an artist was counted over subsequent and perhaps equally great works. Similarly, primary or initial scientific discoveries were given prominence over sustained and later contributions that added vital if not crucial details, albeit not as dramatic. Creative contributions with more of a "splash" in a profession, Dennis claimed, were favored.

Dennis also chastised Lehman for including short-lived creative individuals, thereby reducing the sizes of the samples in the older years. With increasingly fewer creative people alive at ages 50, 60, and 70, compared to a fairly constant number of living 20-, 30-, and 40-year-olds, there were fewer countable individuals and products in the later years. Consequently, tallies of late-life productivity were based on smaller and smaller samples, thereby favoring the earlier over the later years. Dennis met this problem of dwindling sample sizes with increasing age by tallying the productivity of a sample who lived until 80. He found decline, too, although somewhat later than Lehman. (However, this study has its own shortcomings; see Chapter 6.)

For all these reasons, Dennis was skeptical of Lehman's findings. "There is a reasonable doubt that the curves presented by Lehman depict an age decline," and he therefore concluded that Lehman's facts about decline were, in a word, "spurious" (Dennis, 1956, p. 333).

Lehman's work is also criticized by Simonton (1990b, 1991a, 1991b). The objections carry substantial weight because of his extensive research that extends the productivity approach to creativity and supports the decline model. From this authoritative stance, Simonton lists a number of limitations in Lehman's studies, some of which were noted previously: Age curves for creativity vary with the domain of activity; productivity is only one measure of creativity; a decline in productivity is not proof of a decline in creativity; creativity never declines so much as to disappear or become appreciably lessened, that is, decrements are not substantial; secondary peaks in creativity occur later in life, and although not as high as the first one, they suggest a resurgence of creativity in the later years; quality is less affected by aging than productivity, that is, quantity declines more than quality; and when creativity is defined relatively, in terms of "the proportion of hits [notable achievements] to total attempts, the age decrement vanishes altogether" (Simonton, 1991b, p. 627).

Nevertheless, on the whole, Simonton (1991a) supports the decline position, although he raises objections that temper the age-decline connection. "[I]t is possible for creative individuals to antici-pate continued activity right until the end of life…. The picture [of declining productivity] is not as bad as it seems…. The career trajec-tory reflects not the inexorable progression of an aging process tied extrinsically to chronological age, but rather entails the intrinsic working out of a person's creative potential by successive acts of self-actualization" (pp. 14 and 16). In the final analysis, though, despite Simonton's criticisms of Lehman, and in places, arguments and find-ings that support creativity's persistence, he favors the decline posi-tion. Evidently, there is too much confirmatory evidence to overrule it.

But Simonton's work on diminishing creativity with age has not escaped criticism (Csikszentmihalyi, 1989). Romaniuk and Romaniuk (1981) took him to task for relying on post hoc (after-the-fact) analysis, unreliable old date, and concentrating exclusively on well-known figures. Eysenck (1995) added additional objections. Simonton's explanations of decline were "insufficiently robust," he claimed, after noting that 75% of the variance in one of his major data sets was not accounted for. (Variance in this context refers to the usefulness of a single statistic in explaining a finding.) Eysenck also criticized Simonton for excluding alternative hypotheses of decline, especially neurological and personal factors. In addition, he faulted Simonton's explanation of creative potential. It "lack[s] any conceivable psychological support" and although "the theory is inviting, [it is] essentially untestable." Eysenck also criticized Simonton's overemphasis on innate factors, exclusion of individual differences, and insufficient weighing of external circumstances. Eysenck also strongly objected to Simonton's extensive reliance on mathematics. "Of course the curve [for decline] fits observations beautifully, but then all the constants are chosen to make sure of such a fit!" (quotes are from pp. 140–141). He also disagreed with Simonton on other matters, such as the link between creativity and intelligence.

Eysenck's theory of creativity, too, sees creativity as declining with age, but he assigned a major role to personality rather than to cognition or social forces (pp. 138–145). Losses are ultimately due to conditioned inhibition, or more generally, physiology and brain processing. Thus, the decline position, despite serious attacks from different quarters, remains ascendant.

WHERE ARE WE? WHAT'S NEXT?

Declines in late-life creativity based on the productivity of historical figures and the test scores of living individuals have a number of conceptual and methodological difficulties. These have stimulated bountiful objections, strong counterarguments, and cogent rejoinders. Still, the decline position and the measures that support it remain enormously influential and pervasive, in large part because of behavioral and ageistic attitudes that undergird them, as well as the parallels with general findings about aging in which losses rather than gains are emphasized.

But a case can be made for late-life creativity taking directions other than decline. The creativity of many individuals, especially artists, persist into old age; cogent arguments, particularly from the humanities, suggest that creativity has been underestimated in the later years. A good deal more therefore remains to be said, with supporting evidence, that favors other conceptions of the course of creativity (Chapter 5). In addition, Lehman's data on losses in productivity, on which so much of the decline model depends, is open to reexamination (Chapter 6).

Late-Life Creativity

The Continuity Position and Alternatives to Decline

It is always in season for the old to learn.
Aeschylus, *Fragments*

The decline position on late-life creativity has not gone unchallenged. One leading researcher called it a "myth" (Torrance, 1977, p. 16). The continuity model, in which creativity continues beyond the middle years, is the most prominent alternative. It was first formulated around the end of the 19th century, long before the work of Lehman (1953) and the ascendancy of the decline position. Other options come from the humanities, and from anecdotal evidence.

THE CONTINUITY OF CREATIVITY

In 1869, Galton (1822–1911) wrote *Hereditary Genius* (1952), the first major scientific approach to creativity. (Genius is an extreme form of creativity.) Galton and those who followed him (Bowerman, 1947; Bramwell, 1947; Clark, 1916; Cox, 1926; Terman & Oden, 1959) relied on biographical accounts in the historical record rather than productivity tallies. Galton and his successors examined the lives of outstanding figures with highly significant contributions in a variety of fields, including family circumstances, like size and birth order, and from these drew inferences, mainly about the intellect and intelligence of these great men, along with their personal traits. Galton and others concluded that creative individuals were outstanding in intellect, had excellent physical and mental health, were long-lived, continued to add to their original accomplishments through adulthood, and in addition, were highly versatile, that is, displayed extraordinary competencies throughout their lives in areas other than the one in which they initially made their reputations. Thus, White (1931) examined thousands of biographies of 300 men of eminence who lived between 1450 and 1850. He concluded that Goethe was the most versatile individual in history; his talents spanned 18 different fields (of 23 possible), including literature and several other areas of the arts, as well as philosophy, language, history, and medicine; he also had a well-known theory of color. Benjamin Franklin ranked second. Least versatile was Rembrandt, whose biographical record reveals a proficiency "only" in art, although "he would appear more versatile if more were known [about him]" (p. 485). As a group, statesmen were the most versatile; novelists and dramatists ranked second, and poets fifth, tied with scientists; lowest were musicians (ranked twelfth). Artists of all kinds ranked tenth overall, but varied for specific kinds of art: Painters and sculptors were the most versatile, making notable achievements in invention, science, mathematics, handwork, conversation, and administration, followed by novelists, dramatists, poets, and musicians; the latter were versatile in only two areas, art and humor.

In a more circumscribed study of several professions, Raskin (1936) concentrated on the versatility of 120 scientists (biologists, physicists, and mathematicians) and 123 men of letters (poets, novelists, dramatists), again relying on biographies. Like Lehman, she counted the frequencies with which their accomplishments were mentioned. Some of her results paralleled Lehman's: the first works of both scientists and literati were published when they were fairly young, with the former group slightly favored over the latter (average ages were 25.2 and 24.2, respectively); the greatest works were completed about 10 years later, with scientists again slightly older than men of letters (35.4 and 34.3, respectively). However, indicators of versatility occurred later, in the mid- to late-50s (59.4 and 55.0, for scientists and literary men, respectively). Scientists, as a group, too, were more versatile than men of letters, particularly in languages, while the latter excelled in several areas of the arts.

A long life not only provides opportunities for versatility to flourish, but also creativity, according to the Bulloughs and their colleagues (Bullough, Bullough, & Maddalena, 1978; Bullough, Voght, Bullough, & Kluckhohn, 1980). Biographies were examined again in order to study the lifelong accomplishments of notable aging artists of Renaissance Florence and eminent elderly Scottish figures of the 18th century (Adam Smith, Sir Walter Scott, Robert Burns, James Watt, James Boswell, David Hume, and James Mill). Their achievements, the authors concluded, depended on living long: "eminence requires longevity" (Bullough et al., 1980, p. 119).

A long life also benefits the course of scientific creativity. "The common preconception that a scientist's best work is done by the age of 40 and that productivity and creativity decline necessarily thereafter" were rejected as "unnecessary concomitants of aging" (Root-Bernstein, Bernstein, & Garnier, 1993, p. 329). Scientists continued to be productive past middle age because of their many, varied, and changing interests.

[Scientists have a] simultaneous involvement in research in several areas [and] purposely plac[e] themselves in the position of becoming novices again every five or ten years. In effect they become mentally young by starting over again.... Those scientists who are effectively active over the longest time span alter science the most. (p. 341)

Productivity records also support the continuity position. Citation counts to publications, tallied by Cole (1979), showed that aging (60+) physicists, geologists, and chemists stayed active in their fields, as did elderly mathematicians, psychologists, and sociologists. Cole concluded that older professionals "were not much less productive than those under 35" (p. 962), and in addition, the quality of their efforts either became better or did not change with age. Impressed by these studies, Birren (1990) concluded that "There is not a dramatic change [in productivity] over the life span ... The trend [of decline] with age is not large and shows considerable variation within the disciplines." Further, affirmed Birren, "there was no age at which scientific productivity and qualitative output appeared to decline" (quotes are from pp. 29–31; see also Butler, 1990).

Elderly scholars also remain productive. "Senior scholars have ... a normal expectation of productivity...to about the age of 80" (Birren, 1990, p. 33). For architects who were creative in middle age, it continued. Dudek and Croteau (1998) interviewed and tested 70 of the 124 most highly creative architects investigated in a landmark study by MacKinnon (1964). Nearly all (90%), now in their 70s and older, were still working, 61% of them full-time. Importantly, "talent and richness of psychic apparatus continues" (p. 132).

Creativity is also sustained in nonspecialized aging populations. Among 80 women in their 60s to 80s, Rankin (1989) found new or reawakened creative endeavors, with 86% reporting "some sort of creativity activity...at some point in their [later] lives" (p. 168). Creativity included artistic activities, and more broadly, starting new careers. Residents of senior homes, aged 60–93 were observed after three kinds of programs: creative arts senior activities; a home environment where creative artistic activities were carried out; and a nonarts community program for seniors (Dohr & Forbes, 1986). Measures of the participants' creativity in the first two settings increased over time or remained stable. "Older adults do not perceive a decline in creative capacity with age" (p. 123). The continuity position was also informally supported. The host of a radio program on which hundreds of both famous and unknown older Americans appeared as guests concluded that "Creativity in later life was not limited to a few people of exceptional artistic or scientific talent" (Goldman, 1991, pp. 41–42). Creativity in this case was not defined by accomplishments but in more personal terms, with respect to self-esteem and feelings of self-worth, that is, in how a person thinks about himself or herself.

More formal tests of personality also uncover late-life creativity. Responses to certain aspects of the Rorschach inkblot, interpreted as indicative of originality, did not change with age (Andersson, Berg, Lawenius, & Ruth, 1989). The Rorschach, along with other tests of personality, and interviews of pensioners and active individuals about age 70, led Smith (1989) to conclude that "the attitude of creative individuals...towards aging was less negative and their attitude towards illness less defensive" (p. 1). Additional evidence for the positive relationship between creativity and aging are found in Adams-Price (1998b), Goff (1993), Pufal-Struzik (1992), Sasser-Cohen (1993), Simonton (1990b), Smith and Kragh (1975), and Syclox (1983).

The most thorough, rigorous, and well-known studies of creativity's persistence into old age were longitudinal ones initiated by Terman in 1921 (Terman & Oden, 1925–59; see also Holahan, Sears, & Cronbach, 1995; Schneidman, 1989). Terman and his work have influenced the methodology of longitudinal studies, development research and theory, educational practice and special education for the gifted, and importantly, a view of the gifted child continuing to be an exceptional adult well into old age (Subotnik & Arnold, 1994).

Terman and his colleagues initially selected intellectually gifted children whose measured intelligence fell in the top 1% of the population. They have been studied repeatedly into adolescence, adulthood, and old age. The original study emphasized intellect and intelligence, but follow-up studies have also included real-world accomplishments. Four surveys of the original Terman group were conducted in 1972, 1976, 1982, and 1986 (Holahan et al., 1995); the respondents in the most recent survey were in their mid-70s, on average, and were about equally divided by gender (569 men and 494 women); 66% completed all surveys. In the latest study, emphasis shifted from achievements to matters related to retirement and health, and to physical, social, and psychological adaptations to old age. The findings are summarized next.

The original Terman group, now in their 70s and older, show a "second season"; their precocity lasted. Over 1,000 of the original participants, retested as they aged into "later maturity," continued to manifest the positive accomplishments of their earlier and middle years without signs of decline. "The gifted children retained or even enhanced their superiority [especially] in those aspects of living and behavior that rely heavily on language, abstract thinking, and other intellectual skills." They also maintained good physical and psychological health, confirming earlier findings by Galton and his successors.

> The...men persisted in work longer than men in the general population.... [They gave] self-assessments of having lived up to [their] promise [with] positive feelings about work and family [and] feelings of self-esteem and satisfaction in personal, as well as occupational [areas and they have an] enduring interest in meaning[ful] work in the years of later maturity.

The main findings of these monumental studies were glowingly summarized by Holahan et al. (1995):

> In stark contrast to the stereotype that gifted children burn out, these men continue to be productive far into their later years.... We conclude that continuing ability in the occupational realm is possible well into later maturity [and] is surely a contributing force to their quality of life in the later years. [The Terman children] so active and productive in their middle years, have approached aging with continued mastery and competence. For some subjects, occupational achievement, which distinguished this group in their middle years, has continued into later maturity. [The results] point to greater possibilities for the aging years than has heretofore been thought possible.

Indications of their versatility, too, offsets "the stereotype of the gifted as one-sided in abilities and interests [and instead they] have demonstrated a [continued] breadth of interests and associated activities [that reveal] an active group with diverse interests" in intellectual, social, recreational, and cultural areas. (The above quotes are from pp. xi, 20, 80–81, 176–177, and 267.)

Consider the cultural area, which includes creative activities in the arts, such as playing a musical instrument, creative writing, the writing of poetry, and sculpture, as well as less participatory activities, such as listening to recorded music, going to concerts, reading, and attending the theater (Holahan et al., 1995; table 8.1, p. 178, and table 8.8, p. 189). Cultural-artistic activities as a whole were less frequently engaged in than intellectual pursuits, but this depended upon the specific activity. Reading, for example, was engaged in by nearly all respondent (98%), while participation in other cultural-artistic activities ranged from 36–87% of the respondents. Furthermore, activities related to music and art did not decline over the years but showed "moderate continuity." The cultural-artistic activities of this elite group ranked higher than the national average: the men and women in the 1977 survey participated, on average, in about 17 areas, compared to approximately 12 in the general population.

In a less ambitious but also noteworthy longitudinal study, Erikson, Erikson, and Kivnik (1986) asked 29 non-exceptional octogenarians, for whom life histories since the late 1920s were available, to reflect on their old age. They reported "a vital involvement" with life, often maintained by participating in artistic activities. Their involvement, the authors argued, helped them remain relevant, overcome stress, minimize conflicts, lead to balance, and unify their lives.

Support for the continuity position also comes from research on divergent thinking, in which more than one answer to a problem is correct, a feature of creativity that usually declines with age. However, a major study by McCrae, Arenberg, and Costa (1987; see especially p. 136) conceded that losses could be explained by factors other than age, and moreover, they noted large individual differences among respondents of different ages. That is, some older participants were better at divergent thinking than some younger ones. Qualifications on the relationship between aging and divergent thinking were echoed in a review by Mumford and Gustatson (1988), who concluded that declines with age were not always the case, and when found, could be explained in ways other than age.

Mixed results also characterized Roe's (1972) study of the late-life careers of 64 leading research scientists and social scientists aged 47–73. "It is difficult to categorize the … patterns [of productivity over time for they] are quite varied and complicated" (p. 170). In a similar vein, McCrae (1999) argued that creativity persists for some and not others depending on their openness to experience. This explains why "men and women who are outstanding in their fields in their early careers are most often the same one who will be outstanding in their later years" and why some do not fulfill their early promise (p. 365). Evidence for both decline and continuity with aging was discussed by Montagu (1989).

Losses and gains in late-life creativity also characterizes some psychoanalytic accounts of aging. For example, Pruyser (1987) argued that the

> ambiguities of old age and the quality of mourning over actual or anticipated losses may move some old persons toward a late-life creativity that may reaffirm any past creative tendencies…. Although some old people may become more conservative (or even rigid) as they climb in years, aging

> also makes other people more liberal, more free, more lenient, more radical in espousing progressive ideas and causes than they were in their middle years. (pp. 428–429)

Pruyser concluded that there were more positive than negative changes with age. "Older persons show a comparatively greater interiority than ... younger persons or indeed in their own younger years because they have waning Oedipal tensions and conflicts [that] reorganize one's memories" (pp. 431–432).

The poet William Carlos Williams (quoted in Cohen, 1997, p. 5) aptly summarized the inconsistent relationship between aging and creativity: "old age that adds as it takes away." That is, with aging, there can be both increases and decreases in creativity. He may have had artists in mind because of the variety of creative patterns that describe their professional lives. For example, creativity continued into the old age of Michelangelo and Rembrandt, according to Ravin and Kenyon (1998), but not in the case of Piccaso and Munch. Aging has varied affects on creativity, the authors contended, depending on how the artists responded to the losses of old age. If these are overcome, creativity continued or was renewed; if not, creativity disappeared.

Artists may be more resistant to losses in creativity with age than others. In an early survey by Beard (1874), numerous examples of losses in creativity with age were noted for many professions. But the arts were different.

> The most marked exceptions to the law [of decline] are to be found in the realm of imagination; some of the greatest poets, painters and sculptors, such as Dryden, Bryant, Richardson, Cooper, Young, DeFoe, Titian, Christopher Wren, and Michael Angelo, have done a part of their very best work in advanced life. (p. 8)

Beard concluded that the arts are "realms [in which] quite old men have succeeded" (p. 27). Artists are the exceptions, Beard felt, because their reputations and judgments of their work were more subject than other professions to idiosyncratic tastes that changed over time. Why this should lead to a more lasting creativity for artists is unclear. Ironically, Beard evaluated the great painter Turner's last works as "inferior" (p. 15).

Contrary to Beard, Dennis (1955) found that artists' productivity declined more sharply and earlier than other professions (Chapter 4), but Root-Bernstein (1999) challenged this: "People in the arts and literature have *longer* effective careers than the typical person in the sciences" (p. 459; emphasis added). Much more will be said about aging artists compared to other professions in subsequent chapters.

Evidence and explanations in support of the continuity perspective on late-life creativity vie with the decline position. The continuity model also has its problems, many of which are shared with its chief competitor.

PROBLEMS WITH THE CONTINUITY MODEL

The thesis that creativity continues into the later years is supported by empirical research, persuasive arguments, and anecdotal accounts (Chapter 1), and is further buoyed by its optimistic outlook on old age. Nonetheless, it has its problems. Like the decline position, the continuity models not only depends on the vagaries of the historical record (biographies) and the reliability-validity of test scores, particularly of intelligence, but also personality and divergent thinking. Mirroring its counterpart, continuity is also faulted for ignoring actual creative products. Reservations have also been raised about the robustness of late-life creativity: Late-life peaks in creativity are smaller than youthful peaks (Simonton, 1988a). There are also complaints about the overemphasis on personal over cognitive traits, the reverse of the decline position. Objections, too, are raised about subjective inferences on intelligence and personality and the reliance on informal, selective, and incomplete biographical accounts. Equating intelligence with creativity, and IQ with creative test scores, raises hackles similar to those who treat productivity counts as synonymous with creative ability and measurements of creativity as the same as creative potential.

Irritating to some, too, is the genetic undercurrent of the continuity position, especially in its original and starkest form by Galton, the founder of the movement, who held that genius and its cognates—creativity, exceptionality, talent, giftedness, and intelligence—were innate. An emphasis on the inborn or hereditary nature of creativity implies that people are either born with creativity or not, that creativity is fixed, and that environmental influences on creativity are secondary. Egalitarian, liberal, and democratic sensitivities find a nativist orientation offensive, especially when it opposes the belief that creativity can be learned, changed, and influenced by experience.

Bothersome to some "realists," too, is the core belief of the continuity model in the long-term persistence of creativity, which runs counter to the prevailing view of aging as a time of deficits and losses. The charge that the continuity perspective is polyannish is countered by the current optimistic reappraisal of old age, which includes the notion of successful aging and findings that show aging's losses are not as marked or pervasive as previously believed (Chapter 2). The continuity's positive outlook also coincides with recent trends in cognitive psychology and gerontology that emphasize the stability and restoration of some abilities in old age rather than their decline and loss. Thus, Levy and Langer (1999) argued that "The life span developmental [continuity] model is best supported by existing research on creativity" (p. 45). Those who disagree, might counter that the apparent ascending curve of the continuity position describes only the most recent period and not the long-term picture that favors the decline model.

With facts, examples, and arguments both pro and con either decline or continuity, a reasonable option might be to combine both antagonistic views of late-life creativity: Creativity declines for some and persists for others. An eclectic solution makes it possible for two opposing viewpoints to coexist, and results in a synthesis that mirrors the complex nature of creativity (Chapter 1).

However, an eclectic solution postpones, or at worst, brushes aside, the inevitable question: Why does creativity with increasing age decline for some and continue for others? The question is lost if masked by a superficial compromise. The two positions are, after all, based on underlying incompatibilities, such as the innate and learned origins of creativity, and these irreconcilable elements do not disappear when added together. The tensions that underlie each camp may be temporarily set aside, but the inherent contradictions between them will eventually emerge and be difficult to recognize because of their non-partisan gloss.

A shallow consensus also prematurely ends the vigorous promotion of one view or the other by motivated proponents. In addition, a belief in only two options for late-life creativity discourages the pursuit of other possible pathways. More than decline and continuity are likely given the multidimensional, -factorial, and -determined nature of creativity (Guilford, 1950, 1967; see also Schneidman, 1989; Chapter 1), its linear (straightforward) and nonlinear trajectories, scientific and artistic variations, among others, as well as the expressive, technical, inventive, and innovative forms it takes (Taylor, 1959, 1973). Creative expression is also not regular; there are pauses, regressions, and stasis, especially during the period of so-called incubation, when there is no apparent overt activity (Cohen-Shalev, 1989). In addition, late-life creativity can occur gradually or suddenly, evolve slowly over time or emerge suddenly, and these differ in their frequency of occurrence (Bornstein, 1984, especially p. 132). Different manifestations of creativity may have different sources, end-products, and peaks, and take irregular directions, none of which are consistently described by continuity or decline.

The way creativity is manifested over the lifetime, furthermore, may reflect the different demands on younger and older individuals, and these lead to different kinds of late-life creativity (Cohen, 1986; Cohen-Shalev, 1989b; Levin & Kahana, 1967; Nelson, 1928). Certain pressures arise in childhood, these differ from middle-age, and others are unique to old age. The outcomes of each may be manifested in ways that cannot be folded within either a model of continuity or decline. Changes in the nature of conflicts and stresses, how they are appraised, and the way they are resolved also account for the emergence of creativity at one age rather than another. Individuals who become creative in their later years are called "Ulyssean" (McLeish, 1976), after the Greek hero who began his odyssey when he was about 50 and ended his travels at about 70.

The wide variety of possibilities for late-life development makes it unlikely that only two models, either decline or continuity, account for late-life creativity. At least four other logical possibilities exist: Late-life creativity can increase, suddenly emerge, take new forms, and change direction (in style, subject matter, media, or materials used). These multiple possibilities are promoted in the humanities, and especially in art history.

THE CONTRIBUTION FROM THE HUMANITIES

The humanities are known for emphasizing the unique lives and singular accomplishments of creative individuals, including its aging exemplars, and the variety of ways in which creativity is manifested late in life. Scholars in art history and literary criticism are keenly interested in individuals who were extremely creative throughout their lives, lost their creativity suddenly or unexpectedly, and expressed their creativity in unusual ways at different ages. Discussions of creative individuals and particular works in youth, middle, and old age highlight aspects of creativity that are overlooked by blanket averages. The emphasis on unique individuals and works forces scientific investigators to step away from group productivity counts and general test scores, and goads them to pay attention to exceptions that contradict either decline or continuity models of late-life creativity. "Statistical generalizations concerning age-related productivity," Root-Bernstein (1999) contended, "need not apply to individuals" (p. 463).

For scholars in the humanities, biographies, not bibliographic counts, illuminate personal, economic, and cultural factors that facilitate or dampen an individual's creativity over a lifetime. The biography of an eminent person "yields data that cannot be provided by the customary empirical research methods [since it emphasizes] the uniqueness of the events...and of individual human lives [with] a richness of detail" (Howe, 1982, p. 1079). When consideration is given to elderly individuals, it is possible to note some are more creative than many younger individuals, and that late-life differs from youthful expressions. For example, aging academic psychologists were marked by declining publication rates overall, but there was also "substantial individual difference" (Horner, Rushton, & Vernon, 1986, p. 319).

Biographical accounts of aging creative individuals that include feelings and attitudes about growing old indicate how increasing physical and other disabilities were perceived and overcome (or not), describe how creativity changed with age (or did not), and illustrate the ways losses and gains were judged and explained. Personal accounts also suggest why creative efforts in old age slowed down, ended, shifted direction, or followed new forms, as well as how these changes were received by colleagues, experts, critics, and the public. In-depth case studies in the humanities go beyond productivity and other kinds of tallies (publications, citations) and test scores by highlighting particular creative works and singular creative individuals. Otherwise, idiosyncratic paths and kinds of late-life creativity would be submerged by group averages.

Narratives of the lives, works, and times of creative individuals do not replace productivity counts and creativity tests. They are supplements. Spotlights on individuals by scholars give substance to scientific generalizations about late-life creativity, raise questions by alluding to exceptions, confirm with specific examples the age at which peaks, plateaus, and declines occurred, and speculate on the factors that might become an impetus for more formal and testable hypotheses. Discussions in the humanities of the different kinds of late-life change broaden narrow scientific preoccupations. Qualitative scholarly approaches to late-life creativity are a valuable adjunct to quantitative scientific research.

Like empirical excesses, qualitative lapses occur, too, with incomplete, unreliable, and "extravagant and fanciful explanations" (Howe, 1982, p. 1080). But when approached with caution, rich narrative accounts are in keeping with the expansive meaning of the term "creativity." Abstract statistical profiles of the decades in which (anonymously lumped) creatives produced their (collectively unspecified) achievements are illuminated when accompanied by examples of works, discussions from experts, and personal testimonies from the people studied.

Regretfully, humanistic views on the variety of ways in which creativity is manifested throughout the life course have not been incorporated within scientific circles whose horizons are bounded by frequencies and test scores from large samples of homogeneously undistinguished groups. Resistance stems from the two-cultures conflict (Chapter 2), behavioral biases, and ageism (Chapter 4). Consequently, the humanities' insistence on aging as more than an occasion for losses, or at best stability, have not been heeded or appreciated by more narrowly focused researchers in creativity. A closer look at creative old artists illustrates how much these scientists have missed.

CREATIVITY AMONG OLD ARTISTS

Old artists illustrate both continuing and declining creativity, as well as other trajectories. The art historian Rosenthal (1968) maintained that the late works of painters "are as varied and different in character as the masters themselves" (p. 14). Another authority, Feldman (1992), affirms the individuality of old painters.

> A great variety of individual responses [are] given by eminent painters in their old age [and] old age can be a meaningful culmination…in the lives of artists in all fields who continue to practice their art with no loss of talent or inspiration, often breaking new ground late in life. (pp. 3–4)

The variety, according to Feldman, stems from the artists' motivation, confidence, and openness to new learning.

Multiple pathways to late-life creativity are possible because of artists' long-livedness, which gives them more time and opportunities to be creative (Chapter 2). A long working also allows creative expression to shift, be altered, and become refreshed. Goethe remarked, "We constantly perceive in men of superior endowments, even in old age, fresh periods of singular productiveness as they seem to grow young again for a time, [to] feel a new growth" (quoted in Nelson, 1928, pp. 110–111). Relatively few artists ended their creativity early in life (Rodeheaver, Emmons, & Powers, 1998; Rosenthal, 1954): Mary Cassatt (because of blindness), Dali, and according to some, Picasso and Kollwitz. Marcel Duchamp, although he produced very little, is an interesting exception (Pritikin, 1990). His last artwork was produced in 1913 when he was less than 30 years old (he died at age 81). Yet despite a long period of apparent nonproductivity, Duchamp influenced his peers throughout his life. His revolutionary conceptual work, *Large Glass*, 1915–23, subtitled *The Bride Stripped Bare by her Bachelors, Even*, is considered by many art historians as one of the greatest works of the century for its innovativeness, revolutionary theory of art, and influence on contemporary works.

Long-lived artists also leave a substantial longitudinal record of their accomplishments. The art of Rembrandt, for example, "documented almost every year or phase of his life from youth to old age," thereby displaying creative growth and change. His many self-portraits (nearly 100 paintings, etching, and drawings, about 10% of his work) were "the most personal of statements [made] of an artist's life [because they were] made essentially for himself." Rembrandt's self-portraits revealed changes in his creativity at different ages, "mov[ing] from jauntiness, vitality, and even [the] arrogance of youth and early adulthood, to the quiet, introspective, and intensely spiritual experiences of old age" (Rosenberg, 1980, quotes are from pp. 89–90).

Michelangelo's creativity also changed dramatically with age (Rosenthal, 1954). His two *Pietas*, sculptured at ages 22 and 90, represent two significant themes of aging. The earlier one celebrates life and physical and sensuous beauty; the older work, in its rough hewn and indistinguishable boundaries (except for the faces), reveals a deeper sensitivity. Similarly notable changes describe the late works of Cèzanne, Mondrian, O'Keefe, and others (Berg & Gadow, 1978; Byers, 1982; Lambert, 1984), especially those known for an old-age style (Part IV). "Had [Goya] died on his fortieth birthday," Sayre (1956, p. 117), asserted, "few people would now be aware of his name. [The] great legacy of his drawings, prodigious in number and profound in content, comes to us out of the last thirty years." Perhaps in implicit agreement, Goya titled a drawing he did when about 80, "I'm still learning."

The large number and many distinct types of late-life changes among Western artists were categorized by Rosenthal (1954). For example, Monet changed only a little, Homer moderately, and Titian drastically; El Greco and Rubens broadened or concentrated earlier tendencies, respectively, while Hartley developed a distinctive style; Degas and Renoir added a new facet to a largely unchanged style, while Corot stayed the same in some ways and differed radically in others; and Bonnard, Corinth, and Hokusa did their greatest work in old age.

Creativity for some artists suddenly burst forth late in life (Henri Rousseau, "Grandma Moses," Grandma [Tressa] Prisbey, Elizabeth Layton, Alma Thomas). Equally dramatic and profound were radical shifts in aging artists' styles, the "Altersstil" or the old-age style

(Part III). Noteworthy, too, were an appreciable number of masterpieces produced relatively late in artists' lives (Chapters 7 and 8). Whether masterpieces or not, Picasso produced about a third of his work in his 60s and later, Chardon and Man Ray manifested a second peak in creative productivity between ages 60 and 69, and 70 and 79, respectively, and Goya's productivity remained high throughout his life.

Examples of the flourishing of late-life artistic creativity are exhibited at special museum shows of paintings, drawings, videos, performance pieces, poems, and photos by elderly artists. An exhibition by 48 artists in their early eighties (Talley, 1990) reflected "the social, political, and philosophical issues that surround aging [and] attitudes about growing old [as well as] ideas." No one theme of aging predominated. "Emotions...range from optimism to deep-seated fear [and take] a variety of directions [with] no single philosophical or aesthetic point of view [and are neither] assuring and positive [nor] dark and frightening" (p. 3).

Aging artists' reactions to growing old were shown at the Corcoran Gallery of Art in Washington, D.C. in a show entitled "Still Working" (Foote, 1994). There were over a hundred paintings, sculpture, and drawings by 32 older men and women who were "underknown artists of age...in American art...who mov[e in his or her] own direction.... Taking different paths...they experience their age differently [with some artists] doing the best work of their lives." The diversity of their subjects and themes reflects the "freedom and independence that [comes] with age. [The artists] have all given up on being right or wrong.... They have stopped asking why, and they really don't care that much about what other people think [and lack any] concern about outside opinion" (quotes are from p. D4).

Testifying to the triumphant spirit of old artists was a juried competition sponsored by AARP, the largest national organization for the elderly. From almost 9,000 entries, 40 works by artists over 50 were chosen; nine of the award-winning artists were 70 years old or older, including an 86- and a 77-year-old. The exhibit demonstrated the artists' "celebrat[ion of] the idea that creativity is a lifelong pursuit" (quote is from the press release). The winning art went on a nationwide tour of three cities and was published in three issues of the organization's national magazine, *Modern Maturity* (Mullin et al., 1990).

The works of 17 famous historical artists who remained creative in old age were exhibited at the National Gallery of art in 1992 (Feldman, 1992). The artists were Bellini, Calder, Cèzanne, Corot Degas, El Greco, Hals, Homer, Ingres, Magnasco, Mattise, Monet, Pissarro, Rembrandt, Tiepolo, Titian, Tiepolo, and Turner. Surprisingly, Goya was not included, although the title of the show was a quote attributed to him, "I am still learning," a remark made when he was about 80. The Smithsonian also hosted a show by elderly artists (Hufford, 1987).

These exhibits demonstrate the variety of aging artists' reactions to old age and the outstanding accomplishments of old artists. Late-life examples of creativity can also be collected and showcased for professions other than the arts and for the general population as well.

WHEREFORE DECLINE AND CONTINUITY?

The various manifestations of artistic creativity in old age, and their implications for creativity and aging in general, challenge views of creativity where the only options are decline or continuity. More pointedly, the accomplishments of old artists raise serious doubts about the inevitable and pervasive decline of creativity, its minimal presence in the later years, and the impossibility of a late-life resurgence. The range of options also lends credibility to the continuity position, as well as broadens it to include variations of sustained late-life creativity. The multiple possibilities of late-life creativity are provocative, but without empirical support they are not completely convincing. Unknown, for example, are the number of individuals for whom increases in creativity occurred, in relation to the number in whom it declined, continued, or suddenly emerged. The categories of change that describe the kinds and degree of late-life creativity, once established, also have to be quantified. How many kinds of late-life changes in creativity are there? How many individuals fall in each? Categorization of the types of changes with frequencies concretely demonstrates that late-life creativity does indeed follow a variety of pathways, and provides a framework within which to place an impressive array of illustrative examples, stirring anecdotes, and cogent arguments from the humanities.

These goals were accomplished in this book by focusing on aging members of one profession, painters, establishing the facts for this group, and thereby outlining a model for other kinds of aging creative individuals. Studies identified the ages-of-creativity for elderly painters, described their late works, examined personal reactions to old age, and documented the different paths taken by late-life creativity. The success of these endeavors should encourage similar analyses of aging creative writers, musicians, and other artists, as well as the creativity of aged scholars, scientists, inventors, and other professionals, and the elderly population at large. By uncovering the facts about late-life creativity, starting with painters, scholars in the humanities may be spurred to reexamine and revise untested arguments, revitalize debates, and initiate new qualitative inquiries into late-life creativity.

The empirical strategy begins in Part III, with a series of original studies on the creativity of aging painters from both the past and present. Setting the stage first, Chapter 6 reexamines Lehman's seminal investigations of the productivity of historical artists. Was there a decline, as he claimed? Subsequent chapters report on the course of late-life creativity in old painters and the art of their old age, including masterpieces, using productivity tallies, questionnaires, and self-ratings, as well as judgments of late works by both experts and laypersons. Individual artists, too, are examined and compared to general age-trends. Part IV concentrates on the old age style, a unique expression of late-life creativity with important implications for models of aging and development.

PART III

Late-Life Creativity

Historical and Contemporary Artists

I see in you the estuary that enlarges and
spreads itself grandly as it pours in the great sea.
Walt Whitman, *To Old Age*

CHAPTER 6

A Reconsideration of Lehman's Findings

Old age and the wear of time
teach many things.
Sophocles, *Tyro*

The youthful and brief span of creativity was established by and rests largely on Lehman's seminal analyses of historical figures (Chapter 3). No less an authority than Terman, a pioneer in the study of gifted children, declared in the foreword to Lehman's (1953) book, *Age and Achievement*, that this work "deserves to be ranked among the most important contributions...to the literature on genius" (p. v). Ironically, Terman was a proponent of the *persistence* of creativity, rather than its early decline, the position championed by Lehman.

Lehman exhaustively surveyed the ages at which creative works were produced by a wide range of historical figures of renown in at least 10 fields of science and 15 different areas of the arts (especially music and the visual arts). He examined the ages at which eminent persons produced their "best, notable, and most important works," as defined by the number of times these were referred to in authoritative sources. Exemplary contributions by famous individuals were tallied, plotted, and tabulated by age. These profiles indicated, first, that creative achievements were systematically related to age; and second, that creative output nearly always reached its maximum relatively early in life, in the thirties.

Lehman's findings have held up for over half a century because of his thoroughness, explicit methodology, and extensive sampling of hundreds of illustrious historical figures from numerous fields over several centuries. Creativity's youthful rise and fall in the past has been confirmed for contemporary figures and with other kinds of measures, like citation counts and creativity tests. Despite extensive criticism, contrary evidence, and forceful counterarguments (Chapters 4 and 5), creativity's early decline is generally accepted. Lehman's pioneering work was extended by a succession of proponents, attacked by opponents, gave rise to alternative models, spurred a great deal of discussion, much of it contentious, and initiated dozens of studies. Consequently, Lehman's approach, findings, and conclusions, accepted or not, have greatly influenced debates, investigations, and theories of creative development. Much of what is known, believed, and expected about the relationship between creativity and age builds upon or is a reaction to Lehman's work. Any discussion of creativity must therefore take his contributions seriously. A critical review of Lehman's work is therefore in order.

I begin with a reexamination of artists, mainly painters, a profession synonymous with creativity, and more importantly, with careers and creativity that extended well into old age (Chapter 2). If Lehman's findings about the youthful and abbreviated span of artistic creativity were suspect, then doubts would be raised about professions outside of the arts. Uncertainties about late-life artistic creativity also undermines the decline model, lends support to alternate views of the course of creativity, and strengthens optimistic perspectives on old age.

This chapter begins with a general critique of Lehman's studies of artists followed by an examination of 13 sets of data on art that support or refute the decline position. It then compares the productivity-by-age of artists and other professions, and concludes with an analysis of the ages of artists who produced masterpieces (Lindauer, 1998a, 1993b).

A CRITIQUE OF LEHMAN'S STUDIES OF ARTISTS

Scholarship in art history and criticism has been considerably revised in the half century or so since Lehman's work was completed. Major technical developments have led to more accurate dating of works of art as well, and attributions of authorship have also been corrected. For example, the number of works assigned to Rembrandt has been markedly reduced since the turn of the century from about 1,000 to approximately 300. Reputations have also waxed and waned since the first quarter of the 20th century, the time period from which most of Lehman's references on artists were drawn. Thus, Lehman's selection of creative artists and exemplary artistic works depended on outdated historical sources. A contemporary list of creative artists would therefore be quite different from Lehman's, as would the number of works they produced and the ages at which they were done. Similar revisions would apply to other professions.

Another serious problem with Lehman's research is the short life spans of many of the artists in his samples (Dennis, 1955, 1956,

1966; see Chapters 3 and 4). Actually, Lehman did examine a set of long-lived artists—findings that are seldom reported—but the results for this group are problematic (see points #7 and #13 in the following section). Consider the 28 artists Lehman called "the world's greatest" because their work was most often referred to in art textbooks. (This group is further identified in point #4 below.) They painted their masterpieces between the ages of 32 and 36, a span that conforms to Lehman's general view of early creativity. But of the 23 artists for whom reliable birth and death dates were found, eight (35%) died when they were less than 60 years old. For example, Raphael died at 38, Watteau and G. Bellini at 47, and Hobbema 48. Overall, the age of death for this select group of extraordinary artists was relatively early, a little past 60 (63.46, to be exact).

Most of Lehman's tabulations of artistic creativity at different ages were therefore skewed in favor of short-lived individuals, because it is based on an increasingly reduced number of living artists, which diminishes the number of works counted after age 30. With artists dying at ages 40, 50, or earlier, those over 60 were not as well represented; losses in sample size with increasing age obviously reduced the frequencies of creative output among older artists. Since lesser and lesser creative works by fewer and fewer artists were produced at older and older ages, an early peak in creativity would be an artifact of a restricted sample size dominated by youthful artists. It is not as if artists die young, thereby presenting Lehman with an inescapable dilemma. Artists are in fact quite long-lived (Chapters 2 and 4), as Lehman himself recognized (see point #13 below).

To remedy the age bias in Lehman's sample, Dennis (1966) tabulated the total productivity of artists, scholars, and scientists who lived to 79 or older. By including only long-lived individuals, an equal number of individuals was therefore represented at each decade of life. Thus, unlike many of Lehman's samples, there were as many artists and members of other professions in their 20s, 30s, and 40s as in their 50s, 60s, and 70s. In contrast to Lehman, too, Dennis counted total productions, not just frequently cited works. Total productivity is not as coarse and indiscriminate a measure of creativity as it may seem. Highly creative people, as a rule, produce a great deal; and quality is a function of quantity (Simonton, 1994, p. 184). Total productivity is therefore a reasonable approximation of creative productivity.

As it turned out, Dennis's global measure or productivity came close to Lehman's more selective measure: Declines in total productivity occurred relatively early in life, in the 40s, a scant 10 years later than the peak Lehman found for the most creative works. The productivity of artists in Dennis's sample, furthermore, declined earlier and more sharply than it did for scholars and scientists.

These results, however, are not without their own difficulties. Dennis' occupational categories of "artists," "scholars," and "scientists" were composed of highly disparate and variable numbers of subgroups. "Scientists," for example, included biologists, physicists, and chemists. The category "artists" contained seven groups, including 25 dramatists and 4 kinds of musicians, and among the latter librettists authored 176 operas and composers wrote 109 pieces of chamber music. There were also three kinds of writers: novelists, playwrights, and poets; and the number of their works also varied widely. For example, the 25 dramatists produced 803 plays. Productivity averages for the category "artists" therefore rest on very diverse groups with widely disparate tallies. The output-by-age for a particular subset of artists, say composers, might not therefore mirror the pattern for the librettists or some other subgroup of musicians, or the overall profile for "artists" as a group. Members of artistic professions, known for their idiosyncrasies, nonconformity, and uniqueness, are likely to be inconsistent. An anonymous aggregate also makes it impossible to pull out the output of composers and writers at different ages and compare them. The same ambiguities apply to "scientists" and "scholars."

A more serious problem with Dennis's sample of "artists" is the inexplicable exclusion of painters, a puzzling omission in light of their long-livedness (Dennis, 1966, p. 2). Perhaps they were excluded because the dates when some paintings were produced are ambiguous. Paintings can be worked on for many years, not just the year in which they were signed. Further, artists of the past did not sign and date their works. However, at least the latter problem has not existed for at least the last two or three centuries. The failure to include painters in the sample of "artists" throws further doubt on whether productivity in this profession, as Dennis reported it, declined earlier and more sharply than the two other groups. A detailed look at Lehman's findings for artists, which follows, raises more serious problems.

LEHMAN REEXAMINED: PAINTERS, ARTISTS, AND OTHERS

Lehman's (1953) conclusions about the creative decline of painters were based on thirteen sets of data (pp. 70–85, 316–319). Five of these sets, however, do *not* show losses in artistic creativity. Needless to say, these have not been emphasized by Lehman or proponents of the decline position. A closer look at these five questionable findings, along with the eight more frequently cited, discloses several contradictions, inconsistencies, and ambiguities that throw additional uncertainties over the decline position, at least for painters, and by implication, artists in general and perhaps other professions as well.

Eight of Lehman's 13 sets of data indicated that painters' creativity peaked at a relatively early age, and although the 30s predominated, the actual ages varied somewhat, depending on the particular sample of works examined: 32–36, 30–34, and 35–39. The following data for artists fit the decline model.

1. 162 etchings by 27 non-British artists.
2. 53 paintings by 32 artists in The Louvre museum.
3. 650 paintings by 168 artists featured in at least two of 60 authoritative sources. The age-of-creativity profile in this set varied, though, depending on the criterion of selection. (See references to "60 sources" in the numbered paragraphs below.)
4. 40 paintings by 28 artists whose work appeared 10 or more times in 60 sources. Lehman labeled these frequently cited artists as "the best." They included the 23 long-lived artists for whom life and death dates were available and whose life spans are reviewed later (Table 6.1).
5. 67 paintings by 67 artists whose works appeared 5 or more times in 60 sources. Lehman called them "major artists."
6. 86 paintings by 86 artists whose works appeared less than 5 times in 60 sources. Lehman labeled them "minor artists."
7. 61 paintings by 61 artists whose works appeared in all 60 sources *and* who lived to 70 or older. These older artists were not mentioned in Dennis' (1956) critique of Lehman's purportedly exclusive reliance on short-lived artists.
8. 61 paintings by 30 of the "most selective" artists (not defined). These works were taken from a large set of 357 artists who produced 7,600 paintings (see point #11 below) and evidently refer to a different pool of artists than those listed earlier.

These eight sets of data demonstrated an early decline for creativity among painters. But not all of Lehman's data for this profession showed creativity peaking in the 30s. Five indicated later ages. These are rarely, if ever, mentioned by adherents of the decline model. The exceptions follow:

9. 84 works by 66 living and modern American artists. Their creativity peaked between 40 and 44.

10. 1,684 paintings by 305 artists whose works appeared at least once in 60 sources. Remarkably, their creativity *rose* with increasing age, peaking between 90 and 99 (Lehman, 1953, p. 75)! Lehman downplayed this unexpected and startling finding, calling it an "artifact," and speculated that the late upsurge might be due to the inclusion of idiosyncratic artists or the "bizarre judgments of some individual compilers" (p. 77). To downgrade this contrary finding, Lehman revised his criteria, making them more selective. Artists were now chosen whose works were mentioned at least twice or more, up to 10 times. With this adjustment, the data for artists supported Lehman's thesis (see points #3–7 above).

11. 7,600 paintings by 357 artists. In this case, productivity followed a flat (no-change) pattern through at least ages 50–54. Lehman adjusted this problematic set of data by reducing the number of artists and art, and consequently found corroborating evidence for the decline thesis (see point #8 above).

12. 7,032 paintings by 135 artists (labeled as "least selective") and 506 paintings by 152 artists called "moderately selective." These artists' creativity peaked at ages 32–45 and 42–47, respectively (the intervals are Lehman's). The bases for selecting the two sets of artists were not indicated, but presumably depend on some unstated number of times these artists' works were mentioned in 60 sources (see point #3 above).

13. 2,273 works by 302 artists arranged according to the various ages at which they died and the average year of life in which they made their "contributions" (Lehman, 1953, table 59, p. 318). This set of data is especially critical and deserves a more extended examination.

The term "contribution" is not defined, but probably refers to the age at which the most creative (frequently referred to) paintings, literary works, musical compositions, scientific discoveries, and other major contributions were made by the painters, writers, composers, scientists, and members of other professions, including medicine, philosophy, leadership, and athletics. Unlike other analyses, the sources for these samples were cited and were quite extensive. Thus, four volumes were used for artists and art (Champlin, 1914); and five different references were consulted for the sciences. It is not clear if these sources were used in any of Lehman's other analyses.

The samples of artists and works were quite large: 302 painters, 242 authors, and 662 composers contributed 308 paintings, 334 historical novels, and 279 musical compositions. Moreover, they were selected from an even larger pool: 2,273 oil paintings, 757 historical novels, and 2,548 grand operas. The findings are arranged in Table 6.1 for artists (painters, writers, and composers) and other professions (the column labeled "13 groups") from the sciences, medicine, philosophy, leadership, and athletics. Because of the way Lehman presented this data, it was impossible to disentangle the combined data for all professions, including the artists. However, the artists' impact on the combined data should be relatively minor, since the number of painters, writers, and composers were far less than the sample sizes for the 10 other professions. The data were noteworthy in a number of ways, and since they are rarely, if ever, cited, they are reviewed in detail.

The table is extremely illuminating (and complicated, which is perhaps why it is rarely referred to) and requires a close reading. Consider the column "Painters" in Table 6.1. It shows the number of painters represented at several age-of-death intervals, ranging from "below 50" to "80+." The second line of the body of the table ("50–59") is read as follows: 57 painters who died between 50 and 59 made their contribution at age 40, on average. The last line ("80+"), indicates that 33 painters who died at 80 or later made their contributions at age 48. Note that the number of painters who died before age 69 (the first 3 age intervals combined) or after 65 were about equal: 147 and 155, respectively. Consider again the last line of Table 6.1 ("80+") for painters, writers, composers, and other notables from 13 professions. Of the three sets of artists who died when they were

Table 6.1. Age of Contributions at Different Times of Death

| | Mean ages of contributions[a] | | | |
| | Artists | | | |
Age of death	Painters	Writers	Composers	13 groups[b]
Before 50	36 (47)	36 (34)	32 (96)	34
50–59	40 (57)	42 (45)	39 (150)	40
60–69	42 (43)	47 (35)	41 83)	41
65–69	45 (45)	51 (31)	41 (96)	45
70–74	46 (35)	53 (30)	42 (82)	46
75–79	51 (42)	50 (26)	39 (73)	46
80+	48 (33)	55 (41)	45 (82)	48

Based on Lehman (1953), table 59, p. 318.
[a]Number of artists are in parenthesis.
[b]Median of all professions' Means.

80 or older, already noted were the 33 painters who made their contributions at age 48. In addition, 41 octogenarian writers made their mark at age 55; 82 of the oldest composers contributed at age 45; and all 13 groups combined (including the three sets of artists) made their contributions at age 48 (on average). Now look at the earliest age ("before 50"): 47 painters who died before 50 made their contributions at age 36. And so on for the other groups at different ages of death.

These data are noteworthy in several regards:

1. The earliest of the seven age intervals in Lehman's table was "below 50," a fairly late time of life against which to plot creative contributions for a champion of early decline. More surprising, the critical 30s and 40s were omitted, ages when Lehman argued that creativity peaks and declines in most professions. The implication is that too creative few individuals died at ages 30 and 40 to be meaningfully tabulated at those decades.

2. Intriguing, too, is the longevity of the professions represented. Of more than 1,000 artists, about two thirds died when they were 60 and older. Specifically, 63% composers, 66% painters, and 67% writers were 60 or older when they died. Among painters, 84% died after 50 and only 16% before then. The long-livedness of artists is again apparent.

3. The most unexpected finding in Lehman's obscure back-of-the-book tabulation is that contributions generally occurred later than the 30s, the age highlighted in the front of the book and by proponents of the decline position. Consider the trend for painters, which is graphically depicted in Figure 6.1: The later the age of death the older the age at which contributions were made. For example, the two sets of 114 painters who died before 59 (the two earliest intervals combined) made contributions between the ages of 36 and 40, which is in keeping with the decline model. However, the 75 painters who died after 75 (the two oldest age intervals combined) made their contributions between approximately 48 and 51. In short, painters who died at older ages made increasingly later contributions as they grew older. There was a slight decrease for painters who died at 80 and older, although they still made later contributions than those who died before they were 74.

The same pattern—later contributions in the older rather than younger years—generally characterized each of the three groups of artists and all professions combined (Figure 6.2). The exact patterns differ, though, depending on the specific profession and the year of death. In general, though, writers made later contributions than the

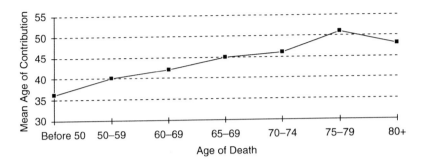

Figure 6.1. Ages at which Painters Made their Contributions as a Function of their Age of Death (based on Lehman, 1953, table 59, p. 318).

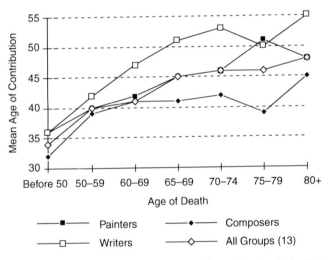

Figure 6.2. Ages at which Artists and Others Made their Contributions as a Function of their Age of Death (based on Lehman, 1953, table 59, p. 318).

other groups at increasing ages of death, composers made relatively earlier contributions, and painters fell roughly between these two groups of artists (except for those who died between 75 and 79). The general pattern for writers and composers, furthermore, seems different from all other professions, while painters appear closer to the other professions.

Thus, early declines in creativity ("contributions") were not the rule when age-of-death was taken into account, but instead, the opposite was true. Artists and others who died before age 50 made their contributions in their 30s, and those who died between 50 and 59 did so around 40, as did the few who died between ages 75 and 79. But by and large, most contributions were made at age 45 and later. Indeed, 48 was the average age at which those who died after 80 made a contribution. For painters, the overall average age of contribution was 44.00 (*Median*=45), an age that is considerably later than the 30s emphasized by Lehman in the first eight sets of data presented earlier. For writers, the average age of contributions was close to 48 (47.71 to be exact) and for composers it was nearly 40 (39.86). These rarely cited statistics indicate that when the age of death of long-lived creative individuals was taken into account, peaks in creativity were closer to the 40s and 50s than to the 30s, which are one to two decades later than the one promoted by Lehman and adherents of the decline position.

Lehman did the same sort of analysis for 980 scientists, recording the dates of their contributions in terms of the decades in which they died (table 58, p. 317). The pattern was the same as artists: late-life contributions were the norm. "The greater the number of decades lived, the smaller the percentage of the sub-group's [of scientists'] total output [was] contributed from 30 to 39" (p. 316). In other words, the early middle years were *not* the most productive for long-lived scientists.

These data are not fatal to the decline position, since the meaning of "contributions" is not clear (which also applies to "most selective" in point #8). Further, eight sets of data did confirm the youthfulness of creativity among artists (points #1–8, above), and for many other professions in different chapters of Lehman's book. Five sets of contrary data (points #9–13) are disquieting, though, especially the 13th data-set, which is a rather large exception. It is impossible to resolve the discrepancies between the two incompatible sets of data (#1–8 and #9–13), since Lehman did not specify the sources upon which nearly all of his data for artists were based. He merely noted that "All of the available art histories in [a] University library were studied" (Lehman, 1953, p. 75). Even if he had cited all his sources, it would not help. As pointed out earlier, the sources he relied upon are outdated and biased by the social, gender, and academic conventions of the first half of this century and earlier, the dates when they were published.

It therefore seems safe to conclude that Lehman's age-trends for artists' creativity are mixed, conflicting, and ambiguous, thereby justifying a reexamination of Lehman's data for other professions. If this future reanalysis of nonartists finds the same inconsistencies as artists, then the problems of the decline model (Chapter 4) are thrown into bolder relief, and alternative models of late-life creativity are encouraged. The groundwork for options other than decline begins in the next section, which establishes the longevity of artists who produced masterpieces, without which it would be impossible to track the long-term course of creativity. These, too, throw more doubts on the adequacy of Lehman's sample of artists.

THE LONGEVITY OF ARTISTS WHO PRODUCED MASTERPIECES

In order to determine the ages at which artists produced their best works, or masterpieces, if it was early or late in life, the sample has to include only the long-lived. If artists as a rule were short-lived, tallies of their productivity would be biased, and the youthfulness of creativity would be favored. There are plentiful anecdotal accounts of the long-livedness of artists (Chapter 2), but it has not been systematically and quantitatively established. Are there in fact a considerable number of long-lived artists? And if there are, does the number of masterpieces—their most creative works—differ for short- and long-lived painters? And for both, at what ages were they produced (Chapter 7)?

To answer these questions, four collections of masterpieces of Western art were examined, a number thereby minimizing the biases of a single compilation. Age 60 was used to distinguish short- and long-lived painters because for artists from the past, 60 *was* an old age; and art historians look for signs of an old-age style at age 60. The number of short- and long-lived artists, the number of masterpieces they produced, and their ages at death, are presented in Table 6.2. (Additional details about these collections are discussed in Chapter 7.)

Consider first the collection by Wood (1986), which contains reproductions of 93 masterpieces by 92 artists (van der Weyden is represented by two examples). The oldest artists were Michelangelo (who died at 90), Chagall (98), Titian (92), and Rouault (88); artists who died young were Masaccio (29), Seurat (33), Modigliani (37), Giorgione, and Raphael (the latter two died at 38). The artists generally lived into their late 60s (dying at age 67.48, on average). Divided into short- and long-lived artists, the former died when they were approximately 46 years old (45.54 to be exact) and the latter at 76 (75.54).

More importantly, there were more long- than short-lived painters, 68 and 25, respectively. Of even greater relevance, the long-lived artists produced nearly three times as many masterpieces as the younger

Table 6.2. Age of Death for Long (60+)- and
Short-Lived Historical Artists of Masterpieces

Sources of masterpieces (and number of artists)	Age at death (Mean)
Wood (1986) (93)	67.48
68 long-lived artists	75.54
25 short-lived artists	45.54
Murray and Murray (1965) (45)	65.44
32 long-lived artists	75.00
13 short-lived artists	41.92
Janson (1959) (138 multiple examples by 31 artists)	67.18
99 works by 21 long-lived artists	75.83
39 works by 10 short-lived artists	49.50
Fisher (1988) (95)	72.05
78 long-lived artists	76.88
17 short-lived artists	49.88
Lehman (1953) (23 best works)[a]	63.46

Based on Lindauer (1993b, table 5, p. 231).
[a]Lehman (1953) obtained data for 28 artists, but the dates for 5 artists could not be determined (Lindauer, 1992). Of the 23 for whom dates were available, 15 died after 60 and 8 before then. The dates of production for Lehman's short- and long-lived artists could not be established because the sources referred to were not specified.

ones: 68, or 73% of the total. (The question of exactly *when* these masterpieces were painted is postponed for Chapters 7 and 8.) A similar pattern characterized the other collections. Take the Murray and Murray (1965) compilation of 45 masterpieces. The average age of death for all artists was about 65 (65.44), but older and younger artists again sharply differed, as expected, dying at 75.00 and 41.92, respectively. The long-lived artists, once more, produced most of the masterpieces (32 or 71%).

The same profile characterizes the Janson (1959) collection, although it sharply differed from the previous two. They included one masterpiece for each artist, or no more than two in a few cases, but the Janson collection contained multiple examples. For example, 31 artists were represented by three or more examples of art. Despite these differences from the previous two collections, most painters in Janson were long-lived: Twenty-one (68%) lived to at least 60 and their average age of death was around 75 (75.83); there were only 10 short-lived artists and they died at about 50 (49.50). As in the previous two collections, most of the 138 works (99 or 72%) were painted by long-lived artists. Of some interest, too, the older artists produced, on aver-age, nearly five masterpieces (4.71 to be exact), while the younger artists painted fewer (3.90), presumably because they had less time to do so. For example, the short-lived Durer and Raphael were repre-sented by seven and five works, respectively, while the longer lived Rembrandt and Michelangelo had twelve paintings each.

The Fisher (1985) collection of 95 works differed from the previ-ous three in that it contained American artists only, whereas Europeans dominated in the others. Appearing on the art scene much later than their European counterparts, American artists lived in healthier times. As a consequence, their overall age of death was later than the European's, about 72 (72.05 on average), compared to about 65–67 for the other three collections. But like the other collections, most of the American artists (81%) died late in life, between 70 and 100, which is considerably later than the Europeans, presumably because of better living conditions. Nevertheless, despite the recency of the Americans on the art stage, and living longer and healthier lives, the pattern was the same as the other collections: Long-lived artists again produced most of the masterpieces, 78, or 82% of the total.

In short, painters of masterpieces were long-lived. Their longevity is remarkable, considering that most lived in harsher times, and when the general population's life span was much shorter. (The artists of Renaissance Italy, too, were long-lived, and moreover, comparable to the life span of contemporary British artists [McManus, 1975].) Perhaps more importantly, the majority of masterpieces (71–82%, depending on the collection) were painted by artists who lived to 60 and older. Furthermore, 68–88% of the world's masterpieces were produced by very long-lived artists, those who died between 75 and 77 (the number varies with the collection).

Compare these numbers with Lehman's sample of the world's "best" artists, the ones most frequently cited in the art history texts he consulted (the last set of figures in Table 6.2; see also point #4). Reliable birth and death dates were available for 23 of the 28 artists, which still leaves a substantial number (82%). Their average age of death was 63 (63.46), which is younger than the range of 65–72 in the four collections surveyed. Further, 65% (15) of Lehman's best artists died after age 60, which is also less than the percentages in the other collections (which ranged from 68–82%). As his critics charged, Lehman did favor short- over long-lived painters, at least with respect to "best works," and compared to masterpieces. The youthfulness of the age at which creativity peaked for at least one set of Lehman's artists, and perhaps others, were biased by their short-livedness.

A long life gives artists more years in which to express their creativity, which seems obvious. But it is also possible that long-lived artists painted masterpieces when they were 30 or so, and thereafter lived on past glories. Thus, the above analysis does not necessarily contradict the decline model. Alternatively, it could be argued that masterpieces were done late in life as death approached, as "swan songs," or due to accumulated experience and wisdom. A third possibility is that masterpieces were produced fairly evenly, throughout the artists' lives. These options are examined in the next two chapters, which tracks the exact dates at which masterpieces and other paintings were produced by historical and contemporary artists.

CHAPTER 7

The Course of Creativity among Historical Artists of Renown

The Peak and Productive Years

> If Providence gives me another two years,
> I believe I shall still be able to paint something beautiful.
> The painter Corot, 1873 (at age 77)

The most creative years for painters of historical renown, according to proponents of the decline position, are the 30s (Lehman, 1953). Furthermore, total creative output declines by the 40s, and earlier and more sharply for artists than for scientists and scholars (Dennis, 1966). Thus, creativity reaches its highest point in young adulthood and falls by the early middle years, a span of only about 10 years or so. Art historians, though, hold that artistic creativity has more than one high point, it is not only in youth, and it is sustained, refreshed, and emergent in old age (Chapter 5). These conflicting assertions about artistic creativity among painters are examined in this chapter. (For details of the research on which this chapter was based, see Lindauer, 1993b.)

The peak years of painters' creativity were determined by recording the ages at which masterpieces were done. Were the worlds' greatest works of art indeed produced when painters were in their 30s, as Lehman claimed? Or later? The second part of the chapter examines the range of years over which painters' creative efforts were sustained. Did their span of creativity decline in their 40s, as Dennis showed? Or later?

THE PEAK YEARS OF CREATIVITY: THE AGES OF MASTERPIECES BY PAINTERS OF RENOWN

The ages at which the Western world's greatest artists painted their masterpieces defines the years of highest creativity. The data for masterpieces were obtained from four collections with the word "masterpiece" in their titles, subtitles, or prefaces. By relying on four rather than one or two sources, errors of omission and biases in the editors' selections of masterpieces and artists were minimized. What makes a work of art a masterpiece is debatable, but specific criteria may not be necessary. The artist and author Fisher (1985) claimed that masterpieces have an immediate self-evident quality about them, "an instant meaning for most people, whether lexicographer, art historian, critic, artist, or the general public" (pp. 8–9). The editor, author, former director of the Old Masters Picture Department of Christie's, and fine art auctioneer Wood (1986) asserted that it would not be too difficult to put together a collection of masterpieces that would "constitute [an] ideal museum" (p. 17). The art historian Janson (1959) maintained that agreement would be found for masterpieces of "historic significance" (p. 1). Nevertheless, differences exist between experts, which is why several collections of masterpieces were surveyed in order to approximate some consensus.

The four collections (also referred to in Chapter 6) differ. The two by Murray and Murray (1965) and Wood (1986) contain mainly European artists from the last few centuries; the former included the "best" or "superior" works that illustrate important aspects of art. The Fisher (1985) collection contains only contemporary American artists, in contrast to the three other sources, which are predominantly European. The fourth set by Janson (1959) illustrated several masterpieces by some artists (e.g., Rembrandt and Michelangelo were each represented by 12 masterpieces) in contrast to one example for each artist in the other sources, occasionally two, but never more than three.

The four collections differed, more importantly, in the number of examples of works and artists. Two collections were similar in number, containing works by 93–95 artists; the other two had considerably fewer: 31 and 45. Consequently, the overlap between the paintings and artists in the four sources was modest; the same works and painters were not in each collection. Altogether, the four collections contained 445 different masterpieces by 120 artists, which suggests more agreement on the greatest painters than on the greatest works.

Of the 445 works, 371 (83%) were dated, thereby making it possible to track their ages-of-production. Undated paintings were mainly from the 17th century and earlier. Excluded, too, were those works with a range of approximate dates that were too wide to be assigned a reasonable midpoint as an estimate. In a few cases where the endpoints of the range were less than 5 years apart, the average of the span was used, thereby minimizing distortions in chronology. A similar approximation was assigned to artists whose dates of birth and death spanned a limited range.

The collections consisted of the following number of masterpieces: (1) The Wood (1986) collection dated 93 of 100 masterpieces by 92 artists (van der Weyden had two examples). (2) The Fisher (1985) compilation contained dates for 95 paintings by 84 artists (Whistler, Shahn, Hopper, and a few others were represented by several examples). (3) The set by Murray and Murray (1965) had a very large number of paintings (1250). To bring the numbers in line with the other and smaller collections, so as not to skew the analysis, only paintings in large page-size formats and in color were tallied. Expensive reproductions, it was reasoned, rather than small black and white examples, were the best examples of the points discussed in the text. On this basis, 45 works by 43 artists were selected (Raphael and Degas had two works each). (4) The Janson (1959) set of 138 paintings contained multiple examples by 31 artists. As already noted, Rembrandt and Michelangeo were represented by 12 works each; Leonardo da Vinci, Donatello, and Picasso had eight examples each. Although the four collections differed from one another in several ways, this did not turn out to be a problem, as will later be shown, since the pattern of results in each case was consistent with the others.

The ages of the artists were categorized as short- and long-lived painters in order to avoid the charge that differences in life spans distorted the years of creativity. Artists who died at age 60 or later were placed in the "long-lived" category. Considering the times in which most artists in these collections lived, the 17th to 19th centuries, 60 was indeed old. Age 60 is also the age at which art historians look for an old-age style (Part IV). Artists who died at 59 or younger were assigned to the "short-lived" category. The average ages at which masterpieces were painted by the long- and short-lived artists, and for both ages combined, are shown in Table 7.1. For comparison purposes, the table also includes the ages at which creativity peaked for the 23 of the world's greatest painters, according to Lehman's analysis of works most frequently cited (Chapter 6).

For all artists, short- and long-lived combined, masterpieces were generally painted in the mid- to late-40s. (Here, as elsewhere, average means are referred to, unless otherwise indicated.) For long-lived

Table 7.1. Average Age (Mean) at which Masterpieces were Painted by Older (60+) and Younger Artists of Historical Renown

Source (and number of masterpieces)	Age when masterpieces were produced (Mean)
Wood (1972/1986) (93)	44.10
68 long-lived artists	47.57
25 short-lived artists	34.68
Murray and Murray (1965) (45)	44.58
32 long-lived artists	48.81
13 short-lived artists	34.15
Janson (1959) (138 multiple examples by 31 artists)	46.41
99 works by 21 long-lived artists	48.03
39 works by 10 short-lived artists	30.55
Fisher (1985) (95)	49.51
78 long-lived artists	51.93
17 short-lived artists	38.38
Lehman (1953) (23 best works)[a]	32–36

Based on Lindauer (1993b), table 5, p. 231.
[a]Five artists were excluded because reliable birth–death rates were not available. See Note, Table 6.1.

artists, though, the most creative paintings were completed later, in their late-40s to early-50s, and for some individuals it was even later. In short, when longevity was taken into account, and this accounted for the majority of artists, creativity peaked long after the 30s, contrary to advocates of the decline model (Chapter 3).

The 30s did stand out as the decade in which short-lived artists painted their masterpieces. But even this age is late relative to the artists' average age of death. Thus, youthful artists painted their masterpieces when about three fourth of their lives had passed. In contrast, long-lived artists did their best work when relatively *less* of their lives had been lived, at approximately the two third mark. In this relative sense, short-lived artists were "older" and long-lived ones were "younger" when they painted their masterpieces. (The age-of-deaths of the artists, upon which the relative ages of productivity were based, were reported in Table 6.1, Chapter 6.)

Each of the four collections illustrated these overall findings. In the Wood (1986) collection, all artists combined produced masterpieces at about age 44. But for the 68 long-lived artists, the age was somewhat higher, closer to 48, and for the 25 younger artists, masterpieces were painted at a much younger age, at about 35. Further, older artists completed about half (53%) of their masterpieces when they were in their late 40s, with another quarter (28%) when they were in their late 50s.

Now consider the relative age at which the Wood masterpieces were produced. The short-lived artists died at about age 46 and their masterpieces were completed at approximately age 35. Taking both figures into account, 76% of their lives had passed when masterpieces were painted. Turning to the older artists, they died at about age 76 and produced their masterpieces at approximately age 49, which means they had lived 63% of their lives when their greatest works were painted, which is, relatively speaking, earlier than short-lived artists. Thus, the masterpieces of long-lived artists, compared to their youthful colleagues, were painted later in absolute years and younger in relative terms.

The same patterns describe the other three collections. The masterpieces in the Murrays' (1965) collection were produced at age 44.58. (If the extremely precocious Velasquez is excluded, whose masterpiece was painted at age 19, the average age increased substantially to 49.77.) For long-lived artists, masterpieces were produced when they were 48.81, or when 65% of their lives had passed. In contrast, the comparable ages for the younger artists were 34.15 and 81%, respectively, or younger in absolute years and older in relative terms, compared to long-lived artists. As in the Wood collection, masterpieces were painted well into the later years of these long-lived artists: By age 50, 53% of the older artists' masterpieces were completed; by age 60 another 22% were painted; and an additional 25% were done in very old age.

In the Janson (1959) collection, the average age at which masterpieces were produced was 46.41. This figure is close to 48.03, the average age at which long-lived artists produced their masterpieces, but much later than the average age, 30.55, for the short-lived artists. The long-lived artists continued to be creative: 82% of the masterpieces were painted when they were in their 60s and 56% in their 50s. The artists with multiple examples mirrored these results: The six artists who painted more than six masterpieces did so at age 44.93, on average; and for 16 artists with four or more masterpieces the age was 47.25. In relative terms, 63% of the older artists' lives had passed when they produced their masterpieces, but unlike the other collections, the relative age of the younger artists' masterpieces occurred at about the same time, when 62% of their lives had been lived. The parity between short- and long-lived artists is possibly due to the inclusion of several masterpieces for each artist, thereby spanning a longer span of time and consequently canceling age-of-production differences.

The Fisher (1985) collection, which included only contemporary American artists, followed the same pattern as the other three collections: Short-lived American artists produced masterpieces at about age 38, which in relative terms occurred when approximately 77% of their lives had passed. The long-lived artists, on the other hand, produced their masterpieces later, when they were nearly 52, but they were relatively earlier, when 68% of their lives had past. The American artists sustained this creative output: 30% of the masterpieces were painted when the artists were in their 70s and older, another 5% were painted by the end of their 50s, and 4% more by the end of their 60s; 61%, though, were done by the end of their 40s.

The similarities in the age-patterns for peaks in artistic creativity across four different collections are impressive. The collections were quite different in their (unstated) criteria for selecting art and artists, the audiences to whom they were addressed, cost restrictions on reproductions, especially those in color, and other editorial decisions that affected content. Despite these major differences, the findings were quite clear and consistent: When longevity is taken into account, masterpieces were painted fairly late in life. For the long-lived artists, averages were in the late-40s to early-50s, and even later for some. In contrast, the short-lived artists peaked in their mid- to late-30s.

The collections were useful for another purpose: they identify the world's greatest (most creative) artists. (The more recent American masterpieces by Fisher, 1985, were excluded because most lack the established reputations of the Europeans in the other compilations.) To assign ranks, the artists were placed into three categories. Fifty-eight, represented just once in the three collections, were of the "third rank." Among them were Courbet, Constable, de Chirico, van Dyck, Dali, Daumier, Duchamp, Ernst, Klee, Kirchner, and Leger. Another 31, in two of the three collections, were labeled artists of the "second rank."

Thirty painters who appeared in each of three collections (26% of the total) can justifiably be called "artists of the first rank." They included Bracque, Caravaggio, Cézanne, Copley, Durer, Degas, Delacroix, Eakins, Goya, El Greco, Gainsborough, van Gogh, Giotto, Hals, Homer, Ingres, Leonardo, Michelangelo, Monet, Masaccio, Pollock, Raphael, Rubens, Rembrandt, Seurat, Tiepolo, Turner, Valasquez, West, and Whistler. Most (77%) were long-lived.

The 30 artists of the first rank were not too different, at least in number, from Lehman's (1953) 28 "best artists," the painters cited in at least 10 of the 60 sources he used. However, the artists in both sets were not the same. The differences reflect Lehman's dated sources, changes in artists' reputations over time, and differences between experts' opinions. Nonetheless, nine artists (32%) were in both the Lehman set and the three collections of this study (excluding again the contemporary Americans in the set by Fisher, 1985): Da Vinci, Gainsborough, Hals, Masaccio, Michelangelo, Raphael, Rembrandt, Turner, and Velasquez. These nine might be called the "giants of art." Their reputations have persisted over a long period of time, have held for different sets of experts, and have persisted through shifts in taste, art history, and scholarship.

THE SPAN OF CREATIVITY AMONG HISTORIC ARTISTS: THE AGES FOR CONTINUED PRODUCTIVITY

We now know with some specificity the ages of highest creativity, as defined by when masterpieces were painted by historical artists of renown, and with longevity taken into account. Peaks in creativity occurred later than previously believed, at least a decade beyond the 30s. (More specific figures are presented in Chapter 8.) What, then, of the duration of artists' creativity, that is, the number of years they continued to do creative work after their peak years? Put another way: How long does creative productivity last? According to the evidence marshaled by Dennis (1966), reported earlier (Chapters 3 and 6), creative productivity generally declined in the 40s, and for artists it was sooner and sharper than scholars and scientists. Thus, the generally accepted span of years between the peak and declining years of renowned Western painters is rather short, dropping off drastically after the most creative years. Or is it the case that, like the revised peak years, creative productivity occurred later, lasted longer, and remained high?

To answer this question, the total artistic output of both historical and contemporary painters were tabulated for individuals who lived to at least 80. This insures that young, middle, old, and very old ages were equally represented at each working decade. The criterion was met by 70 artists in one standard text on art history (Hartt, 1985). Only 21 (30%), though, had large and dated collections of their work. Seven artists were represented in "Catalogues Raisonnes," which are complete collections of an artist's output. For the remaining 14 artists, fairly complete collections of two kinds were consulted. Five artists were represented in books about their lives that included extensive examples of their work at different years; these are called "representative" sources.

Profuse examples of the life's work of an additional nine artists were in exhibition catalogs; these sources are referred to as the "best works." The three kinds of collections naturally differed in their completeness. The seven Catalogues Raisonnes contained more than 18,000 works, averaging 2,590 examples for each of the seven artists surveyed; the prolific Picasso's total was 12,603. The representative sources for five artists contained much fewer examples, over 2,000, or 464 for each artist. The exhibition catalogs for nine artists included 1,277 examples, or 142 works for each artist. Although the three kinds of collections different in absolute number, their patterns over different decades were similar when converted to percentages. Consider the three artists whose works were represented in more than one kind of collection. Duchamp's work was displayed in a Catalogue Raisonne and an exhibition catalog; Sloane's art was in an exhibition catalog and a representative source; and Ernst was in the three kinds of publications. While the number of works for each artist differed considerably, depending on the source consulted, the works produced at each age in the different sources were highly correlated, at least .90, depending on the artist. (Correlations of .30 are acceptable in the social sciences.) These findings have practical importance: it is not necessary to consult a Catalogues Raisonnes to determine an artist's productivity over time. This shortcut is fortunate, since Catalogues Raisonnes are not published for most artists (they are too expensive).

For the three collections as a group, the 21 artists produced nearly 900 works each (average=883.10). Even when the highly prolific Picasso was excluded, these numbers are underestimations. Two of the artists, Chagall and Tobey, were still alive at the time their works were tallied, so that paintings produced by them after the sources were published were not included.

The 21 artists represent an array of different styles, nationalities, and eras (Lindauer, 1993b, table 1, p. 222). Their working lives, for example, spanned several centuries, although most (15 or 71%) painted primarily in the 20th century. Nine Frenchmen and eight Americans were included. Common to all, as planned, was their longevity. Their ages at death ranged from 80 to 100; and four were over 90 when they died. Their average age of death was the mid-80s (Mean = 86.50, Median = 87.50).

Age trends in productivity were obtained by tracking the percentage of an artist's output over eight decades of life, beginning with adolescence ("<19") and ending in the 80s and later (">80"). The productivity of the 21 long-lived artists are graphically displayed in Figure 7.1. The percentage of paintings produced through the

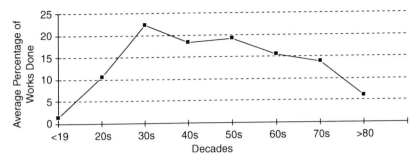

Figure 7.1. Productivity of 21 Long-Lived Artists (Based on Lindauer, 1993b, table 2, p. 223).

seventh decade are based on the total number of works done over seven decades; these add up to 100% (taking rounding-off into account). After age 79, though, the percentages were based on the artist's productivity over eight decades, thereby insuring that the averages were not distorted by the very small number of works produced by a limited number of artists at the oldest decade. The main analysis, with the most data, focused on age 79 as the cut-off point. In calculating the date when a work was produced, the artist's *next* birthdate was used, rather than his last one. For example, if an artist died after his 79th birthday but before his 80th, a work was considered to have been done in the eighth decade. The age at which an artist died, for purposes of consistency, was therefore also calculated as occurring on his or her next birthday. Thus, artists who died after their 79th birthday were considered to be in the eight decade of their lives. To have done otherwise, using the example cited above, would have meant that a work produced by an artist who died after his or her 79th birthday would be recorded as produced in the artist's seventh decade, an impossibility. Hence, artists' ages are one year older than some other sources might indicate.

Three decades were most productive: the 30s, 50s, and 40s, in that order, when approximately 23%, 20%, and 19% of the artists' works were produced. Creative output between this middle period between the 30s and 50s was essentially the same, or flat. (The small percentage differences between decades were not statistically significant.) The periods of lowest productivity were the artists' youth, the 20s (when 11% of their paintings were done) and old age: the 60s (15%) and 70s (14%). The very young and very old therefore produced less than the middle-aged, but note that these extremes did not differ appreciably from one another: very young and old artists were equally productive, and the latter actually produced somewhat more (although not statistically different).

The productivity of the seven artists in the Catalogues Raisonnes and the nine in the "best works" category were separately analyzed. (The five artists in the collections of "representative works" were excluded because the sample was too small to reflect a reliable pattern; the productivity of one idiosyncratic artist would markedly skew the results.) The results for each of the two sets of works were consistent with the overall findings: productivity remained high and fairly similar through the three middle decades.

Creative output did decline appreciably for most artists when they were very old, past their 80s. But seven artists stood out: Braque, Chardon, Duchamp, Hals, Peale, Sloan, and Trumbull produced 8% of their life's work in the 80s, which is not much less than the 11% produced by all young artists as a group. Another six very old artists produced a respectable 6% of their total output at 80 or older: Benton, Ernst, Goya, Matisse, Rouault, and Titian. The decline at this very old age is exaggerated, though. Fewer artists were alive after 79, reducing the number of works in the eighth decade that could be recorded, compared to other periods. In addition, two of the very old artists, Chagall and Tobey, were still alive when collections of their work were published, thereby excluding their recent work. Additions from more recent compilations of works by Chagall and Tobey would therefore increase the productivity numbers at the oldest decade. Thus, the lower percentages in the 80s were less reliable than the numbers for the other decades, and moreover, were probably underestimated. They should therefore be cautiously interpreted.

A more serious problem at every age interval was the nearly complete absence of women artists; only one was included, Georgia O'Keefe. Not many women artists from the past are discussed in the standard texts, perhaps because of sexism. Further, of the few who are, not many lived to 80 or had extensive publications of their work. The situation differs for contemporary women artists, but it will take some time before they live to 80 and collections of their life's work are compiled.

These caveats, however, should not detract from the main finding: For 21 long-lived and famous artists for whom there were fairly extensive records of their life's work, creative productivity continued beyond their 30s and 40s into late-middle age (the 50s), and to a modest degree, for some, to very old age. Renowned artists did not stop doing creative work once past their youth, and while there was some decline, it was gradual, in the 60s and 70s, and most marked in the 80s, although the extent of the losses at this period is uncertain because of the low numbers. Sustained artistic productivity, along with the relatively late age at which creativity peaked—the dates for masterpieces, discussed in the first part of the chapter—make it clear that creativity's best years have been considerably underestimated, as has its span and continuation over time.

An extended period of artistic creativity supports the continuity model and raises more questions about the decline position. Declines occurred at an early age for some artists, later for more, and only in very old age did losses occur, and these were modest. The abilities that underlie artistic creativity, like a vivid imagination and a rich imagery, to name just two, are sustained into the later years. A similarly optimistic view of old age should hold for nonartists.

Keep in mind, though, that the results and interpretations apply to painters, especially great ones of historical renown who are no longer alive. Fortunately, future research can fill in the gaps for contemporary painters, other kinds of artists, like musicians and writers, as well as for scientists, scholars, members of other professions, and laypersons. The strategies of research reported in this chapter can be easily applied to the creative peaks and productive spans of any group: Record the years in which the most noteworthy works were done and the decades over which they were sustained.

The question addressed in Chapter 8, though, is more modest: Does the span of creative productivity differ for men and women artists? Recall that O'Keefe was the only woman whose creativity was examined. The research in Chapter 8 makes up for this shortcoming by including a large number of women artists. In addition, the size of the sample was more than doubled. Further, in keeping with the emphasis in the humanities on taking individuality into account, the creative productivity of individual artists was compared to artists as a group.

Creative Productivity, Gender, and Individual Differences for Long-Lived Artists of Renown

Our nature here is not unlike our wine:
Some sorts when old continue brisk and fine.
Sir John Denham, *Of Old Age*

The creative achievements of 21 historically well-known and long-lived painters continued through middle age (Chapter 7). The small sample's size, and lack of women (except for one, O'Keefe), raise questions about its representativeness. Consequently another 24 long-lived artists were added to the original sample of 21, all with Catalog Raisonnes (complete collections of their life's work). The enlarged sample of 45 long-lived artists encompassed 64% of the 70 long-lived artists discussed in an art history text (Hartt, 1985), and of these 45, 31 (69%) had Catalog Raisonnes. The artists and their nationalities, birth and death dates, the age at which they died, the total works they produced, and the number and kind of sources that were consulted in tallying the number of works they did in different decades are covered in Lindauer (1993b, table 1, p. 222).

The chapter addressed another major complaint of art historians and by scholars in the humanities, namely, that scientific generalities based on group data ignore individual differences. In response, the productivity patterns for specific artists were compared to the general trend. Also met was the criticism that the original study of 21 artists included only one woman.

THE PATTERN OF CREATIVE PRODUCTIVITY
FOR 45 LONG-LIVED ARTISTS

The number of works produced by 45 long-lived artists was traced over eight decades. A clear picture emerged (Figure 8.1): Creative productivity between the 30s and 50s was fairly equal, high, and higher than both the earlier and later decades. (The productivity-per-decade for each artist, all artists combined, and for men and women separately are found in Lindauer, 1993b, table 4, p. 227.) Specifically, 20% to 24% of the artists' works, as a group, were produced in their middle years, compared to 8% in their 20s, 11% in their 70s, and less than 2% at the two most extreme decades. The inverted-U pattern mirrored the results for the smaller set of 21 long-lived artists surveyed in Chapter 7 (Figure 7.1), although the middle years for the larger set of 45 is more regular because of the larger numbers involved. Importantly, the most productive decade was the 40s, a peak 10 years later than the 30s, the time period promoted by the decline position.

The two youngest and oldest decades (<19 and 80+), though, should be cautiously interpreted. The number of living and working artists in their 80s was quite low (13, or 29% of 45) and with one exception they produced very little then (1–3% of their life's work); the exception was DuBuffet, who produced 8% of his work in his 80s. The small number of artists and works at this late age reduced the reliability of the numbers at this age, compared to earlier time periods when all 45 artists were alive. However, Bacon, DeKooning, Chagall, and Tobey were still alive and producing at the time their latest works were published. Since their late and latest works were unavailable, productivity for these four artists is underestimated. Comparisons

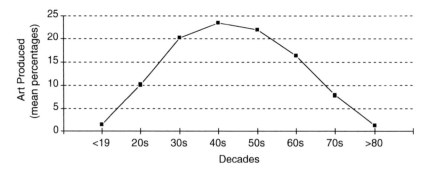

Figure 8.1. Productivity for All Artists Combined.

between the 80s and earlier time periods were therefore somewhat ambiguous.

The adolescent years were also problematic; most artists produced very little in their teens. Even the prolific Picasso is credited with only five works at that time. Yet there were two notable exceptions: Vigee-Lebrun produced 15% of her work in these early years, and Duchamp produced 18 works before he was 19 (17% of his total). Consequently, like the extremely old years, the youngest years were not as reliable as the other decades.

GENDER DIFFERENCES IN THE CREATIVE PRODUCTIVITY OF LONG-LIVED ARTISTS

There were fewer women artists of record in the past than men, and fewer Catalog Raisonnes of their work. Women artists, furthermore, have the same problems as all artists—their work has to be original and marketable—and additional ones: few models to emulate, a marginal status in society and the art world, and unique responsibilities of childbearing and -rearing. These career handicaps are often not recognized. "Art historians have not specifically addressed old women as a category of artists shaped by the aging process, sex, or [the] social construction of gender [although] they have noted a number of demographic, institutional and cultural issues that affect women's abilities to be artists across the life course" (Rodeheaver, Emmons, & Powers, 1998, p. 208). Older women writers, too, are neglected because they publish less than men (Wyatt-Brown & Rossen, 1993). "Growing older affects [women's] literary creativity [and] individual writing careers begin to change in middle age" (p. 2). These gender differences in the arts are not reflected in leisure activities, though. "Women have been…more involved in reading and passive forms of leisure [than men] as well as in sociable, expressive, and artistic endeavors [and women] are more likely to engage in cultural activities" (Cultler & Hendricks, 1990, pp. 173 and 175; for forms of creative expression taken by women in their sixth to eighth decades, see Rankin, 1989).

Broad discussions of women artists (Dunforth, 1989; Pettys, 1985) and of creativity among women, include such issues as sources of encouragement and forms of expression (Reiss, 1999), but aside from an extensive study of older creative women artists' personalities (Orwoll & Kelley, 1998), there is a paucity of research on women artists and gender differences in late-life creativity. "Investigations of gender differences in adult creative achievement," Baer (1999b) asserted, "have been relatively few in number...limited in their focus or quite speculative (and sometimes polemical)" (pp. 753–754). Baer pointed out the anomaly of finding more gender differences with tests than for actual creative performance, where "studies have tended to find no significant gender differences" (p. 756). Nevertheless, Baer concedes that "The consistent lack of gender differences in creativity, both in creativity test scores and in the creative accomplishments of boys and girls makes it difficult to build a strong case for innate gender differences in creativity" (p. 758). Despite the paucity of evidence and their inconsistencies, Baer concluded that men and women artists differed. "There is a considerable amount of evidence of gender differences in creative accomplishment in a wide variety of fields, including the arts" (pp. 753–754).

A concerted effort was therefore made to include as many long-lived women artists as possible in the sample. To this end, 12 women artists were included, although it was difficult, and a few compromises had to be made in the criteria of selection to include even them. Thus, Hepworth, known primarily as a sculptor, was selected, as were the short-lived Labille-Guiard and Morisot, both of whom died in their 50s. To balance these exceptions, three men artists were also included: Delacroix, who died relatively young, in his 60s, and Bernini and Donatello, who are primarily known as sculptors. These few exceptions, as will be shown, did not affect the age-by-productivity patterns in the larger set of long-lived painters.

Gender differences in creative productivity were examined by matching the 12 men and 12 women artists as closely as possible in nationality, age of death, and time periods in which they worked (Lindauer, 1993b, table 3, p. 226). Matching minimized, as much as possible, differences that might arise from a host of factors irrelevant to gender but reflective of the societies and time periods in which the artists lived. Several matches, however, were unavoidably imperfect because of the small number of artists to work with, especially women, and with respect to nationalities. Three cases were particularly strained: The Morisot/Leger pair diverged sharply in longevity, and the Vallayer-Coster/Peale, and Kauffmann/Goya pairs differed in nationalities. The 24 artists represented, for the most part, two broad time periods: Fourteen worked primarily in the mid-19th through the 20th centuries, and seven lived from the mid-18th to the mid-19th centuries. The remaining three time periods spanned the 14th to mid-18th centuries.

The matched women and men artists were as follows (women are listed first): O'Keefe/Benton, Cassatt/Sloane, Nourse/DeKooning, Hepworth/Bacon, Laurencin/Bonnard, Labille-Guiard/Moreau, Morisot/

Leger, Vallayer-Coster/Peale, Vigee-Lebrun/Delacroix, Kauffmann/Goya, Kollwitz/Ernst, and Munter/Weber. The mean age of death for the men artists was 80.10 (Bacon and DeKooning, in their 80s and still alive at the time their productivity was measured, were excluded from this calculation). The women artists' mean age at death was lower, 74.08, but the difference was not statistically significant.

The most productive decades for both men and women were again the three middle decades (Figure 8.2), an inverted-U-shaped pattern for the two sets of artists that is consistent with the profile for the larger set of 45 artists (Figure 8.1). However, there were some notable gender differences. The three most productive decades for the women artists were the 30s through 50s, but it was later for the men, the 40s to 60s. Put another way, women outproduced the men earlier in life, but men outproduced the women thereafter. In their 40s and 50s, though, there was little difference between the productivity of men and women artists. Perhaps because of their earlier outburst of activity, women artists also tended to outproduce the men over their entire lifetimes, doing about a 100 more works, on average (means were 430.42 and 332.08, respectively), although the difference was not statistically reliable.

The gender difference was confirmed by a statistical technique based on ranks that is statistically appropriate for small samples. For this analysis, the extreme and problematic time periods, adolescence and the 80s, were excluded. Since most artists produced very little at those extreme ages, their exclusion would not distort the results. Five time intervals were therefore plotted, from the 30s to the 70s. The previous findings were mirrored: Women's productivity ranked first in their 30s, an age when men ranked third, while the reverse was true in the 50s, when men ranked first and women third.

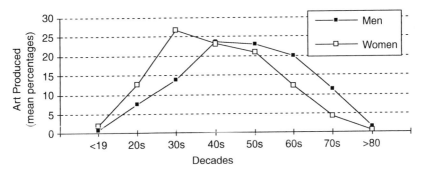

Figure 8.2. Productivity for Long-Lived Men and Women Artists.

Women produced less in their later years probably because of sexist biases that hampered their work, reputations, sales, and exhibiting opportunities. Raising a family, and in some cases, supporting a husband, often an artist, too, also reduced women artists' time and energy for painting in their later years.

The shift in women and men's productivity rates should be cautiously interpreted, though. A few idiosyncratic artists in a relatively small sample can disproportionately affect the results. Labille-Guiard and Morisot died when they were younger than the others, and Korn began to paint when she was in her late 50s. Further, Labille-Guiard, Vallayer-Coster, and Vigee-Lebrun produced most of their work by the end of their 30s (72%, 66%, and 82%, respectively). As already noted, some of the matches between men and women on ages of deaths and nationalities were strained.

Further research on aging women artists' creative productivity over time awaits the availability of a larger sample, which should occur as sexist prejudices are reduced and more women become artists, live long, become famous, and have Catalog Raisonnes published. Until then, gender differences in the arts, like elsewhere, and as indicated by the literature review earlier, remains a complicated issue. At least for the moment, though, it seems fair to conclude that the creative productivity of long-lived men and women artists differed.

But the similarities between genders were also noteworthy: Both men and women artists had long spans of creativity, extending considerably beyond their 40s. The sub-sets of 24 matched men and women artists, like the larger set of 45 artists, were productive through their 30s to 50s, and at a fairly high level beyond the 60s (and the 70s for men).

These findings were consistent with two additional analyses of the data: (1) The productivity-by-age pattern of the 20 men artists tracked in Chapter 7 did not differ from the 12 men who were added in this chapter; and (2) the 32 men did not differ from the 12 women. Reassuringly, too, the spans of productivity for men and women were consistent with the relatively late age at which creativity peaked, that is, when masterpieces were painted (Chapter 7). Later peaks and longer spans, past the artists' 40s and not in their 30s, are in accord with the continuity rather than the decline position.

Summing up the Last Three Chapters. The level of creativity among long-lived artists of renown was quite clear: It was high and fairly constant through mid-life. The 60s and 70s were relatively less productive than the middle years, but as productive as the youngest periods, the artists' teens and 20s, if not slightly more than the earliest one. Creative productivity among painters continued into the later years rather than declined at an early age. Individual artist's creative output, reviewed next, also indicated that creative productivity can be sustained late in life.

THE CREATIVE PRODUCTIVITY OF INDIVIDUAL ARTISTS

Nearly all scientific studies of creativity (or any subject) depend on aggregates and averages, whether of artists or non-artists, productivity counts or test scores, peaks or yearly output. Studies for over half a century, from Lehman to Simonton, have tallied hundreds if not thousands of creative accomplishments by dozens if not scores of creative individuals. The research reported here is no exception, although some mention was made of Picasso in Chapter 7, and in this chapter, Chagall and other extremely long-lived men, along with a few unusual women artists.

Art historians and artists have vociferously objected to this homogenization of individual differences in scientific studies of creativity (and elsewhere), voicing their skepticism about universal statements that apply to everyone. Each artist and each work of art, critics of generalizations and quantification argue, is distinct from other artists and paintings; averages are empty if not distorted abstractions. For scientists, though, generalizations and quantification are valuable for smoothing out unique and irregular instances that are exceptions to the rule. Resolving the issue of the general vs. the individual is a major stumbling block to interdisciplinary exchanges between the sciences, humanities, and arts (Chapter 2).

Averaging uncovers general trends in creativity, but masks the fit, if any, between particular artists and the norm. A compromise between an overly rigid and too loose an approach contrasts the output of individual artists against the framework of artists as a group. How do the peak years of Picasso compare with painters in general? Does a particular painter who contributed to the overall pattern conform to it? Or is the overall pattern an artifact of combining dissimilar individuals? Are the 30s to 50s the most productive decades for particular artists, as it was in general? Or are there as many different patterns of creative productivity as there are artists?

To resolve these questions, the productivity pattern for each of the 45 artists who were collectively summed was visually inspected.

No single pattern fit all artists, but there were not 45 individual profiles. No one decade was the most creative—but there were not 45 different peaks (Lindauer, 1993b, table 2, p. 223 and table 4, p. 227). The details follow.

Thirteen artists (29%) painted most of their works in a single decade—but it was not the same decade in each case. Chagall and Duchamp peaked in their 20s, when they did 28% and 40% of their work, respectively. Chagall, for example, produced 9–18% of his work in decades other than the 20s. For Ernst, Trumbull, and West, the 30s were the most productive, when they did 30%, 41%, and 23% of their life's work, respectively. The 40s were most productive for Balla and Donatello, who produced 41% and 43% of their work then, respectively. The 50s was the high point for Dubuffet (39%), Hepworth (32%), and Moreau (46%), while peaks occurred in the 60s for Bernini (24%), Bonnard (46%), and Korn (56%).

Seven artists (16% of the total) were about equally productive in two decades, but like the single-peak artists, the specific time periods varied and they were not always adjacent to one another. Delacroix's most productive decades were his 20s and 50s and for DeKooning they were the 40s and 60s. Most active in his 50s and 70s was Man Ray. Carra, O'Keefe, and Sloane, though, were most productive in their 30s and 40s while Titian and Tobey had adjoining peaks in their 50s and 60s.

In summary, nearly half of the artists (20, or 44%) had one or two highly productive decades, and these spanned a wide range of time periods, from youth to old age. Despite these variations, a familiar pattern emerged: Slightly more than half of the 29 single and double peaks in productivity (15, or 52%) occurred relatively late in life—between the artists' 50s and 70s. For the other high points in productivity, nine occurred at the end of the artists' 30s (31%), in keeping with the decline model, and five took place by the end of the 40s (17%). Put more strongly, the highest levels of creative output for most artists (69%) took place after their 30s, with the average (median) peak in the fifth decade. Thus, the later years stood out more than earlier ones for individual artists' productive years, just as they did for artists as a group.

A similar pattern of relatively late creative activity characterized secondary peaks, defined as decades in which a relatively large percentage of an artist's works were produced that were at least 10% lower than the primary peaks but considerably higher than the rate at other time periods. Admittedly, the visual assignment of primary and secondary peaks in productivity is subjective, and disagreements between different observers is possible. As a check on the reliability of these judgments, and to minimize any bias, the peaks were assigned on three occasions, with the first two separated by several years. The three sets of judgments were reasonably stable: There were no changes in the assignment of primary peaks (the highest percentage) and the assignments of secondary peaks were consistent with few exceptions.

A total of 90 primary and secondary peaks were observed, an average of nearly two per artist (Mean = 1.98); most (58, or 64%) were primary peaks. About two thirds of all peaks (67%), primary and secondary, were produced in the artists' 40s or later; only one third occurred by the end of their 30s. Looked at slightly differently, more than half of the artists (58%) had primary and secondary peaks by the end of their 40s, 79% by the end of their 50s, and 84% by the end of their 60s.

The peaks in productivity for a few artists were exceptional. Matisse had two primary peaks, in his 20s and 40s, when he did 27% and 24% of his works, respectively, followed by several secondary peaks. In his 30s, 19% of his work was completed and in his 50s and 60s he did 20%. Matisse was therefore fairly equally productive in every decade of his long life. Leger stood out too, with four primary peaks: in his 40s (when 23% of his life's output was done), 50s (22%), 60s (22%), and 70s (18%). Nine artists had three primary and secondary peaks: Bacon, Degas, Laurencin, Munter, Nourse, Peale, Picasso, Vivee-Lebrun, and Vallayer-Coster.

Summing Up. Bursts of creative activity varied for 45 well-known artists; peaks were found in nearly every decade of their lives: in youth, middle-age, and old age. Despite differences between individual artists, creative output generally occurred relatively later in life rather than earlier; and creative productivity continued into old age in nearly all cases. Youth is therefore not the only or even the predominant period in which creative productivity was maximized.

THE SUSTAINED NATURE OF CREATIVITY

Creativity's persistence over artists' lifetimes was affirmed across several different studies, samples, and analyses (Chapters 6–8). The findings held for painters as a group as well as for individual artists, and for singular masterpieces, total productivity, and output per decade. The relative lateness of creative productivity characterized complete and partial collections of artists' works, and both men and women artists. Numerous studies therefore converge on this one point: Creativity persisted into late life.

Artists' creative efforts therefore peaked relatively later than previously believed, in mid-life and beyond, rather than primarily or exclusively in young adulthood; and they were sustained longer, until at least the late-40s and early-50s, if not later in some cases. The span of creativity, from its peak to decline, is therefore at least 10–20 years later and longer than the 10 years between the 30s and 40s, the range championed by proponents of the decline position. Artists do not simply live long (Chapters 2 and 5), but use this extended time to be creative. These conclusions depended on a sample of long-lived artists. To have excluded septuagenarians and octogenarians, and failed to distinguish their accomplishments from younger artists, would have underestimated the years in which artistic creativity peaked and continued.

The individual analyses of 45 artists demonstrated, more than the grouped data could, the varied ages of creative activity. Scholars are therefore quite right in insisting on the importance of considering particular artists. But scientists are also correct in seeking generalizations that summarize a great deal of disparate information. Both molecular and molar perspectives were integrated by comparing individual accomplishments against the group as a whole, a context that was more revealing than focusing on either the group or individual approach independent of the other. The expanded time over which creativity takes place, as well as the individual patterns of creativity for specific artists, corroborate the continuity position (Chapter 5) and positive views of aging (Chapter 1). There are therefore good reasons to be optimistic about maintaining abilities into old age.

At this point, though, unbounded optimism about long-term creativity is premature, for it assumes that the creative patterns for aging painters would also hold for all artists. Would creativity also peak later and continue throughout life for old sculptors? Elderly architects? Aging musicians? Other questions remain. For example, why did some artists deviate from the general pattern?

But before research can be expanded into these other areas of the arts, and eventually, to other professions and the population at large, more needs to be known about aging painters. They are a base upon which to build inquiries into other fields. Take contemporary aging painters. Do they follow the same pattern of late and sustained creative productivity as long-lived and deceased artists of renown from the past? This is the question addressed in the next two chapters.

CHAPTER 9

Contemporary Old Artists on Their Late-Life Creativity

The Quality and Quantity of Late-Life Art

> Father Time is not always a hard parent, and,
> though he tarries for none of his children,
> often, lays his hand lightly on those who have
> used him well.
> > Dickens, *Barnaby Rudge*

An impressive number of well-known and long-lived men and women Western artists from several centuries and representing various nationalities remained creative well beyond their youth and middle years (Chapters 7 and 8). Masterpieces of art were produced in the artists' 40s and later, and artists remained productive until at least their 50s, with many continuing into their 70s and up to their deaths. Artists of renown, in short, exhibited relatively late peaks and long spans of creativity. A few of Lehman's (1953) studies of historical artists, in contrast to those generally favoring declines, also supported the continued tenure of creativity (Chapter 6) as did some studies of creativity test scores among nonartists, especially longitudinal efforts (Chapter 3). Late-life creativity therefore takes two contradictory trends (Chapter 3), either that of decline or continuation. Opposing trajectories could be the result of comparing test scores of nonartists with real-world achievements of artists, or of contrasting cross-sectional measures with longitudinal achievements over time.

Different patterns of late-life development may also characterize contemporary artists. Although the creative reputations of living old artists may not last, there are good reasons for studying them, foremost among which is they can be asked direct questions about their creativity. A qualitative approach, utilizing interviews and questionnaires, rather than tallies of creative works or test results, is endorsed by the humanities. It provides first-hand, individual, and personal accounts of the ways aging and late-life creativity are perceived to influence one another. Accordingly, 60- to 80-year-old men and women artists reported on various aspects of their late-life creativity. These could corroborate declines in late-life creativity, paralleling creativity test scores of living individuals, or disclose continued creativity, as masterpieces by old artists from the past did. The critical question is this: Does the span of creativity among today's aging artists begin as early and last as briefly as the decline position would have it, or is it as late and as long as renowned historical painters displayed, as proponents of the continuity model would maintain?

The study of reports by living old artists is another way of exploring the relationship between artistic creativity and old age. The results for a single profession are also of general interest. If aging artists are able to paint, this demonstrates the persistence of late-life cognitive, motivational, and other abilities, like problem-solving, persistence, and goal-setting, along with their translation into high levels of performance. Why should not the same abilities exist among aging nonartists?

THE STUDY

An open-ended questionnaire on changes in late-life creativity, together with rating scales, were sent to approximately 160 graphic artists in three age groups: the 60s, 70s, and 80s (Lindauer, Orwoll, & Kelley, 1997). The artists were nominated by a random sample of 154 artists who were over 60 years old and listed in *Who's Who in American Art* (1990) who personally knew "particularly creative" individuals. The nominators used a 7-point scale to rate the nominees' creativity, as well as the characteristics that led to their choices and how well they knew the nominees. The nominators also rated the nominees' intelligence, wisdom, and happiness and described their personal and other traits on a 300-item adjective check list. Thus, the selection of older creative artists was based on a great deal of information and certainty.

The nominees completed the questionnaire at their leisure. Up to 88 artists (55% from the original set), about equally divided into men and women, completed the questionnaire; not every question was answered by every artist, and some replies were not sufficiently clear to be scored. The results were combined for both men and women because gender differences were infrequent.

The questionnaire probed several of creativity's elusive character-istics (Chapter 1) and the results are reported in this and several subsequent chapters. The artists typically wrote a variety of rich and complex reports that contained several reasons for change. However, only those reported by the greatest number are presented here. The written answers to the questionnaire were assigned to nearly three dozen categories and quantified through a formal procedure known as content analysis by the three coauthors of the study, independently; discrepancies were resolved by their consensus. Other characteristics of creativity probed by the questionnaire are presented in Chapter 10 and Part IV. In this chapter the focus is on the quality (creativity, orig-inality) and quantity (productivity, output) of the artists' work. Had these changed with age, in what way(s), and how? Following their written replies, the artists consulted a 10-point scale to rate the quality of their work over 10 decades, spanning the past, present, and future; the same scale was applied to quantity. Useable replies for quality and quantity were received from 79 and 73 artists, respectively.

The artists' comments on the quality and quantity of their work over time, and other aspects of creativity covered in subsequent chap-ters, were assigned to one or more of 35 reasons for change. Thus, the same categories were used for several aspects of creativity. Conse-quently, the reasons for change were not independent of one another; an artist could use more than one to account for one change with age and the same reason for several kinds of changes. Broadly speaking, changes referred to the artist, the work, and external (or situational) circumstances. Changes in the artist with age included 11 possible reasons, like increased knowledge; shifts in the artists' work over time could be attributed to 16 reasons, such as developing new techniques; and possible external-situational reasons for change numbered eight, like decreased family responsibilities. The actual reasons given are specified in the discussion of the results that follow.

THE QUALITY OF LATE-LIFE ART

The quality of the artists' work improved with increasing age for each of the three age groups; ratings on the quality of work were high-est for the 60- and 70-year-olds (Figure 9.1). Artists in their 80s were somewhat of an exception. The quality of their work did not improve after the 60s, but flattened, and by their 80s, a slight decline had set in. Nevertheless, their later years were rated higher than several earlier

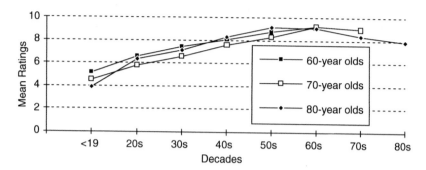

Figure 9.1. Quality Ratings Over Time (from Lindauer, Orwoll, & Kelley, 1997, figure 1, p. 136).

time periods. An 85-year-old woman paraphrased Matisse: "The last [painting] you did is always the best." A 60-year-old woman humorously explained her continual improvement over the decades this way. "I'm expecting diminishing capabilities, and since that may be a self-fulfilling prophesy, I work on overcoming that." A 60-year-old artist was looking forward to getting older, "I haven't yet painted as well as I could and as close to my image of what I strive for."

Nine reasons accounted for most of the gains in quality. The dominant one, reported by nearly a third of the artists (32%), attributed increasing quality to greater knowledge, which included new learning, developing skills, and more experience, practice, and training through reading, study, and trial and error. "I got better because I studied the works of my contemporaries and peers," wrote an 83-year-old man. "I increased my involvement with art, so I became more skillful and knowledgeable, and increased my aesthetic standards," commented an 80-year-old man. A 65-year-old woman replied, "I continued to study and to learn—from life and nature."

A substantial but fewer number of artists (16%) reported that changed circumstances increased the quality of their art, the second major reason offered. These referred to finding new opportunities and ventures, exploring different kinds of materials, increasing availability of resources and facilities, and meeting the demands of gallery owners. Changes were also due to different modes and styles of living, and most simply, having more space in which to work. A 68-year old women summed it up this way: "The environment can affect one's art positively or negatively." But then she added, "Your work depends more on what's outside your head than what's in it!"

A third relatively important reason for qualitative improvements with age was having more time to work, reported by 15% of the artists. They had retired, family duties were reduced, and children had grown up, were out of school, or moved away. An additional reason for growing quality, reported by a modest number of artists (13%), was giving their work a higher priority. This shift occurred because of improved finances and economic freedom, resurgent religious feelings, and reduced civic responsibilities. Some artists added that priorities changed because of new relationships and an increased awareness of what was happening around them in the world at large.

Another relatively frequent reason for change, also noted by 13% of the artists, was a greater acceptance of themselves and their work. Artists had come to terms with what they were trying to do and the ways to do it. "I have a surer hand," was the way a 65-year-old man put it. The artists also said they felt more comfortable about making mistakes, and indeed, expected to do so, and did not feel bound by conventions or by what others said was "good art." Related to self-acceptance, the artists also reported a greater sense of accomplishment, self-confidence, trust about themselves, and faith in their abilities and skills. "I can do better than Picasso," wrote a 71-year-old woman. "If I don't believe I am his equal, how can I create?" A 63-year-old woman expressed her self-confidence this way: "I now feel free to do what I want in painting. After many years of trying to do salable, 'relevant,' or so-called important work, I feel free to be myself and not concern myself about how others might judge me or my work."

In a confident vein, too, 11% of the artists wrote that a greater acceptance of others and the human conditions improved the quality of their work. They had a greater awareness, discernment, understanding, and appreciation of people in general. Along related lines, the artists admitted to being more sensitive to others, hinted at in this 83-year-old woman's comment: "My art has become more intense and I think more profound due to a greater understanding and appreciating of the human comedy."

Only 11% of the artists indicated that physical factors affected quality, a reason that ranked seventh among the top nine reported. The artists admitted that energy had waned and strength had declined, but insisted that adjustments compensated for losses, thereby preventing the quality of their art from being hampered. For example, a 68-year-old artist wrote that rather than face the rigors of the outdoors she worked inside her home more often. Compensating for physical losses, too, was increased motivation, reported by 10% of the artists. "If I do not develop further it will be because I cannot will it. I have willed my development since my fifties," wrote a 76-year-old man, and then added, "I have a desire to express myself, a conscious challenge to grow." Related terms used by others were enthusiasm, commitment, ambition, "being hooked," and "searching for a way." The last of the most prominent reasons offered for increased quality was maturity, mentioned by 8% of the artists, who also alluded to increased growth and development.

THE QUANTITY OF LATE-LIFE ART

The quantity of the 60-, 70-, and 80-year-old artists' work also increased steadily over the years for each age group (Figure 9.2), just as quality did. Like quality, too, the highest ratings occurred late in life. For artists in their 60s and 70s, the sixth decade was the most productive period, just as it was for quality. A 65-year-old woman explained her sustained performance this way: "I take a positive approach to life and my life. I believe in myself and what I can attain through drive, enthusiasm, and hard work." For artists in their 70s and 80s, though, productivity flattened or declined slightly after their 60s, relative to earlier periods, just as quality did. Although output did not increase in very old age, it was still high and higher than most of the earlier decades, including the 30s and 40s.

Seven reasons accounted for increased output by a fairly substantial number of artists. Six of these did double duty: They also explained increased quality. Thus, 14–16% of the artists referred to new knowledge, changed circumstances, and increased motivation as leading to gains in quantity. Physical factors, which also affected quality to a

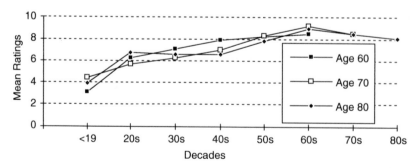

Figure 9.2. Productivity (Quantity) Ratings Over Time (from Lindauer et al., 1997, figure 2, p. 137).

relatively minor degree, again played a peripheral role in productivity; it was mentioned by 14% of the artists. Changed priorities, previously noted as increasing quality, also boosted productivity, according to 19% of the artists. "I'm no longer chasing $$$ or pursuing love affairs," said a 63-year-old woman. "If either arrives, fine; if not, I'm happy this way [as an artist]." A 65-year-old man explained how changed priorities over the years led to more productivity:

> In my teens and twenties, I was preoccupied with dating, mating, earning a living, etc. In those early years, I lacked a focus on artistic production. In my 40s, 50s, and 60s, [changes in] time, interest, and focus increased my artistic production.

Of the six reasons that accounted for increases in both quality and quantity, having more time most often explained the latter; it was reported by a third of the artists. Three comments are illustrative:

> I have never really had time to work until now. In the past, I worked at night and any odd time I had; there was no real development of a theme or a series of works. (Reported by a 71-year-old woman)

> The more time available for creativity, the more artistic productivity. (A 79-year-old male)

> In my teens, I had studies other than art. In my twenties, I was at art school and teaching. In my thirties and forties, I did full time teaching. In my fifties, sixties, and seventies, I was doing full time painting. (An 80-year-old woman)

Only one explanation of gains in quantity failed to be a major explanation of increased quality: Decreased family responsibilities. It was mentioned by 23% of the artists (and only by 5% as affecting quality). Not having to raise a family or worry about children meant artists could spend more time on their art. A clarification from a 76-year-old woman: "Art demands a lot of time, and when you are working and raising a family there isn't much free time." Several artists were quick to point out, though, that freedom from family responsibilities did not mean isolation from family, or indeed, from other social groups. In some cases, new responsibilities were substituted for family responsibilities, such as volunteering.

QUALITY AND QUANTITY COMPARED

Both the quality and quantity of artists' work, according to self-ratings and written replies, persisted into old age. Moreover, with some exceptions, it increased through the 60s and 70s. The two aspects of creativity, furthermore, were closely related. Correlations between quality and quantity ratings were very high in each of the three age groups, ranging between .90 and .99. In addition, six of the seven major reasons for changes in quality and quantity were the same. The close association between quality and quantity is further justification for substituting productivity for more subtle measures of creativity (Chapter 3).

However, quality was differentiated from quantity in one major way: The artists rated the former higher than the latter. The distinction was astutely described by a 66-year old man. "I paint less, in time and intensity. But I think it is better; the quality level is higher." A 71-year-old man was more specific. "In sheer physical production, I simply work about 10% less and hence do 10% less."

Summing Up. Older artists reported that their creativity, whether expressed in terms of quality or quantity, or measured by ratings or personal reports, continued into their later years, up to their 80s. The span of creativity is therefore once again shown to be later than the 30s and 40s, continuing into the 60s and beyond, decades that are not only later than the ones championed by adherents of the decline position, but also beyond the best years associated with famous historical artists (Chapters 7 and 8).

Contemporary artists, like the general population, live longer and can therefore experience a longer period of creativity than artists from preceding centuries. Today's artists have a higher standard of living, better medical facilities, greater financial security, more places in which to exhibit their work, and a larger pool of buyers (Chapter 7). Living longer and better, contemporary artists have more time for creativity to continue. It remains to be seen, though, whether any of the nearly 80 artists who completed the questionnaires will eventually achieve a reputation as high and as stable as the masters of world art.

However that turns out, the continuing if not increasing creativity of this highly selective sample of living artists failed to support the decline model of creativity, for whom the 30s and 40s are the watershed of creativity. For aging artists of considerable repute, creativity neither peaked in the 30s nor declined by the 40s, but instead, continued into the 60s and later. While the 80s showed a reduced rate, it was still higher than the younger years.

For artists to remain creative into their 70s indicates that a wide band of cognitive and other abilities were still operating at a high level in old age, and that the motivation, emotions, and actions needed to translate mental and personal talents into works of excellence were not lost or diminished with time. Becoming old was therefore not necessarily a time of losses in thinking, productivity, and usefulness.

Increases in creativity, furthermore, were fueled by a relatively small number of rather straightforward reasons. Prominent among them were increased knowledge, in the case of quality, and having more time, with respect to quantity. Physical explanations of declines in quality and quantity played a relatively small role and in any event were compensated for.

The ascending arc of creativity well into old age is proof of the viability of late-life cognitive development, corroborates a continuity model of creativity, and supports an optimistic stance toward old age. Gains in creativity with increasing age indicate that late-life development is not solely a story of inexorable decline. Criticism of the shallowness of scientific procedures was also muted by the study's reliance on a questionnaire, with its emphasis on qualitative first-hand reports by living old artists. The placement of quantitative ratings within a quantitative framework should make them more palatable to humanists.

Creativity follows multiple pathways: some show losses while others display continuity and gains. Scholars in the humanities, and art historians specifically, will not be surprised. Chapter 10 covers two other aspects of late-life creativity among contemporary artists' reports: changes in their ideas and in their general approach to art. Is it the case that the creative process and style, like quantity and quality, are augmented with age?

CHAPTER 10

New Ideas and Approaches by Aging Artists

What Has Been Learned about Late-Life Creativity?

No wise man ever wished to be younger.
Swift, *Thoughts on Various Subjects*

The 60–80-year-old artists of some repute who completed a questionnaire on the quality and quantity of their work (Chapter 9) also responded to two other kinds of changes in their creativity with age (Lindauer, Orwoll, & Kelley, 1997): (1) in the sources of their ideas, or the "creative process"; and (2) the approach taken to their work, or style.

CHANGES IN THE ARTISTS' CREATIVE IDEAS WITH AGE

A large majority of 88 artists (81%) reported that new ideas emerged with increasing age. "I have definitely become more creative as the years go by, working harder at my art and spending more time with my peers 'brain busting,'" said a 79-year-old man. "My ideas come more quickly as I age," wrote a 76-year-old man.

The artists gave seven major reasons for these gains. Cited most often was self-acceptance (reported by 24% of the artists). A 65-year-old woman wrote, "I am at ease now. I never worry that I won't succeed; things always work out. As long as I have faith and patience, I trust myself and go forward armed with this." Self-acceptance was expressed somewhat differently by this 70-year-old man. "I used to look for inspiration in the work of others. Now I have no patience for that. I want to enjoy and appreciate the work of others, but for my own work, I listen to my inner voice." Self-acceptance also meant

self-confidence, as this 71-year-old woman indicated. "I've been intim-
idated because I started to paint in my 40's—and I'm still defensive
about that late start, but I am gaining in confidence."

Ranked second in importance as a font of creative ideas was
increased skill and knowledge (reported by 17% of the artists). "My
ideas changed as I have progressed in the study of anatomy, psychology, and elements of design," wrote an 80-year-old woman. "I have
become more sophisticated, partially due to years of experience and
partially from taking classes to improve my skills," commented a
61-year-old woman. "I know more now than I knew as a young man,"
said a 69-year-old man, "so I see great progress in my work." An additional source of new ideas was the artists' increased understanding of
themselves and other people (mentioned by 14% of the artists). These
three reasons (greater self-acceptance, increased skill and knowledge,
more understanding of self and others) were also prominent in
accounting for changes in the quality and quantity of the artists' work
(Chapter 9).

Four additional explanations of new ideas, though, were less
frequently reported as affecting creative thinking. First among this set
was a reduced concern with criticism (reported by 14% of the artists).
The artists evaluated their work in terms of personal goals, looked
within themselves for sources of satisfaction, and were consequently
less career driven. The artists became more self-directed, or as one
artist wrote, now strived to "be themselves." A 66-year-old woman
expressed her independence this way:

> I've learned more concepts and ideas and thoughts; and my automatic
> responses have become more trained. I've learned to "grab" things as they
> pop in from my unconscious and not reject them as impossible or ridicu-
> lous or stupid. I had to put simple "me" first, before my critics and my
> super-critical self.

Another impetus for new ideas resulted from shifts in subject
matter (according to 12% of the artists). For a 85-year-old woman, this
occurred as follows. "As a student and early on, I studied nature. This
gradually developed into a love of space and its divisions, and to represent them as beautifully as I could do it." Other artists reported that
changes in subject matter inspired new ideas, but were divided on
whether their work as a result became more realistic with age or
abstract (impressionistic and less pictorial).

New techniques led to new ideas, too (according to 10% of
the artists). A 60-year-old man explained late-life changes this way.
"My art has become less intuitive and more pictorially logical."

A 67-year-old woman explained that working with collage led to a different way of thinking.

> My collages are pasted bits and pieces from my other works, plus images from art history, etc., all fused into one solid work, but sometimes torn apart a dozen times until the essence remains. My paintings go through several different images, each made in the process of finding what seems right. I seem to be searching for the soul of the work.

Discovering new colors, materials, and media, treating light in different ways, and varying the brushstroke were other techniques that led to new ideas with age. "Through years of work devised to solicit a greater sensitivity to design, composition, and color, I have improved [my thinking]," said an 81-year-old man, adding, "Creativity is a cumulative process." New ideas also came from physical limitations that changed techniques. "I do an impression, rather than a careful and exact painting or drawing, because my lack of manual dexterity makes it difficult to do an exact copy." The 80-year-old woman who wrote this added, "But I also don't really enjoy doing that." The last major reason for changed ideas, reported by 10% of the artists, was the result of becoming more experimental. With increasing age, the artists were more willing to try new things, to test different possibilities, and to develop varied rather than stereotypical ways of working.

In Summary. New ideas for improving the art of old age occurred because of increased self-acceptance, greater knowledge, and a deeper understanding of oneself and others. These factors also increased the quality and quantity of art. Less often mentioned in accounting for changes in quality and quantity, but appearing now for the first time as a source of new ideas, were a reduced concern with criticism, a shift in subject matter, the discovery and development of new techniques, and greater experimentation.

CHANGES IN THE ARTISTS' APPROACH TO ART WITH AGE

The artists were also asked about their general approach to art, or changes in style with age. Like their evaluations of aging's impact on the quality and quantity of their work and the sources of ideas, most of the 88 artists (76%) reported positive changes in their approaches. A 63-year-old man wrote: "My approach is more thorough or 'finished.' I don't feel stuck in any particular way of working, except on rare occasions, and then I am often stimulated to start something anew. There are slumps from time to time, but the general direction is forward."

The artists' reasons for changes in their approach, as for other questions, were complex, and touched on several reasons simultaneously. When unraveled, five accounted for the majority of changes; most were also prominent in explaining the three other aspects of creativity reported earlier. First among these was increased knowledge (noted by 19% of the artists). (Increased knowledge was also either first or second in explaining changes in quality, quantity, and ideas.) An 83-year-old man explained how his approach to art changed through knowledge. "I have done an intense study of the works of masters I love and who I always look to for guidance and direction. Since my 50th year I've incorporated Old Master realism, impressionism, expressionism and abstractionism into my personal statement." In a similar vein, a 69-year-old man reported that he had "completed his studies of the old masters, the 17th century classics, and the Dutch, Spanish, and Italian schools," but nevertheless felt "a continual process of study was necessary. Technically, I can never know enough." A 69-year-old woman simply wrote, "I'm better at everything."

Another relatively important influence on style was a change in technique (reported by 18% of the artists), and it, too, was frequently mentioned in answers to previous questions about creativity. In the words of a 60-year-old man, two major technical changes affected his approach to art. "I've just recently decided to use straight oils, no black oil. It's too complicated and limiting. I also now paint directly on canvas with no preliminary sketches." As was true of other accounts, technique was combined with several other reasons for change, such as self-acceptance, courage, freedom, and deliberateness. The interrelatedness of artists' comments is illustrated in the following two quotations by men in their 60s:

> Occasionally, I have flirted with a traditional realistic style, which I can do very well. But that's only been when for some external reason, such as a lack of acceptance or understanding of what I am trying to do in abstraction, got to me. The style I continue to come back to is more abstract each year.

> My work has become more pictorially definite. My concept of pictorial space is richer because of the visual experiences I have gained in museums, intellectual experiences from reading, and the interplay with a few artist friends.

The artists' comments about technique were more extensive than other explanations, perhaps because it is more concrete than other reasons for change and therefore more easily expressed in writing.

Changes in style were also due to increased maturity (for 13% of the artists). Typically, maturity was combined with other reasons for change, such as having more time and altered circumstances, as the following quote by a 65-year-old man indicated.

> There have been various stages in my work. My best works have been the last ones. Maturity, time to concentrate, and a peaceful way of life have all contributed to my growth and progress.

A modest influence on stylistic change was the artists' increased acceptance of themselves, their work, and their creative ability (mentioned by 12% of the artists). Like most of the artists' replies, acceptance was combined with several other reasons, such as increased self-criticism, more confidence, greater selectivity, and a different sense of time.

The last of the relatively prominent reason for changes in the artists' style was increased motivation or drive (reported by 9% of the artists). The following comment by a 67-year-old man suggested how motivation overcame the physical limitations of age and at the same time inspired new approaches to his art. "I used to work outside in all kinds of weather. Now I can't take it, it's physically too exhausting. Instead, I use color slides as references and work in my studio for the most part. Also I work much more slowly and with greater deliberation."

In Summary. Most artists reported changes in the way they approached their work with age, and it was for the better. Of the five most frequently reported reasons given, most often mentioned was increased knowledge, which dominated accounts of other changes in late-life creativity. Prominent, too, was technique, which like other explanations was not given in isolation, but closely tied to self-acceptance, courage, freedom, and deliberateness. Style changes also came about because of increased maturity, often combined with having more time and finding oneself in new circumstances. Also mentioned by a modest number of artists was an increased acceptance of themselves, their work, and their creative talents. Acceptance, too, was tied to other less frequently cited reasons, like increased self-criticism, more confidence, greater selectivity, and a different sense of time. The fifth and last major reason for changes in the artists' creative style was increased motivation or drive, which also compensated for physical losses.

Unlike quality and the two other aspects of creativity reported earlier, explanations of changes in the artists' approach did not reach a high degree of consensus; none accounted for more than 19% of the artists' replies. In addition, there were fewer prominent (high frequency) explanations (five), and these were usually combined with other reasons. The concept of style is an illusive and subtle one, as most art historians would agree (Schapiro, 1953), and was no doubt a difficult phenomenon for artists to capture and express in writing. In contrast, changes in technique with age were the most abundant and unambiguous, probably because it is so specific and concrete a characteristic of art.

THE FOUR ASPECTS OF LATE-LIFE ARTISTIC CREATIVITY

Elderly artists reported that four aspects of creativity—quality (originality), quantity (productivity), sources of creative ideas (the process), and approach (style)—improved with age. A variety of reasons for these changes over time were offered, and these varied in kind and frequency, depending upon which aspect of creativity the artists were asked about. In order to gain a broader perspective on changes in creativity with age, the four sets of reasons were combined. To accomplish this, each of the major reasons for changes in quality, quantity, ideas, and style were ranked according to the number of artists who chose it, and these ranks were averaged across the four dimensions of creativity. On this basis, there were 15 prominent reasons for increases in creativity.

Of the top five reasons for change, foremost was increased knowledge (see Lindauer et al., 1997, figure 3, p. 142). (Its mean rank was 2.00; the lower the number, the more artists mentioned it.) Knowledge had the greatest impact on quality, and to a lesser but still substantial extent on quantity, ideas, and style. Self-acceptance was a distant second (rank = 4.38). It mainly influenced the emergence of new ideas, and to a lesser degree, changes in style and increases in quality; it had little impact on the quantity of work. The remaining three of the five most prominent reasons for change were considerably less often reported than the previous two. Physical changes (rank = 7.25) affected mainly the quality and quantity of work; available time (rank = 7.75) primarily impacted quantity, followed by quality; and new circumstances (rank = 8.35) influenced quality and quantity about equally.

Four additional reasons for changes with age were distinguished from the top five, forming their own subset (see Lindauer et al., 1997, figure 4, p. 143). In this group, motivation (rank = 9.88) rather uniformly affected style, quality, and quantity. The artists' understanding of themselves and others (rank = 10.13) fairly evenly influenced quality and style, too. Changing priorities (rank = 10.88) was important for quantity and quality. The last of this group of five was technique (rank=13.00), which was central in changing style and ideas but less critical for quality and quantity.

The last six of the top 15 reasons for change influenced only one of four aspects of creativity. In this low profile group, maturity (rank = 11.75) mainly affected style; and concern for excellence (rank = 12.75) was a source of new ideas. Falling sharply below these two were the following four reasons for change: A willingness to be more experimental, which led to new ideas (rank = 20.75), which was tied with a shift to either more abstract or realistic works, and whose main impact was on style; reduced family responsibilities (rank = 24.75) affected quantity; and exploring different kinds of subject matter (rank = 25.38) resulted in new ideas.

Thus, a relatively small number of reasons, no more than 15 and perhaps as few as five, accounted for most of the positive changes in four aspects of late-life creativity. The reasons varied, depending on which of the four aspects of creativity they primarily accounted for, many were related to one another, and some were more effective than others. Furthermore, they were neither so abundant nor arcane as to resist systematic analysis.

A rather clear picture thereby emerged about 60–80-year-old visual artists' views of late-life creativity: The quality and quantity of their work improved with age, their general approach to work or style deepened, and the sources of their ideas expanded. Aging therefore enhanced rather than stifled artistic expression. Changes were for the better as the artists grew older primarily because they had learned a great deal, not only about their art, but about themselves, others, and the world in which they lived and worked. Building on accumulated knowledge, the artists improved their technical skills, did better art, and grew personally. That is, they accepted what they were doing along with who they were. Further, their art benefited as artists became more concerned with and aware of the people, events, and world around them with age. Aging enhanced the artists' work in other ways. Their art became more mature, as might be expected with increasing age, but in addition, they gave it more time and attention because of decreased responsibilities from family and job. Altered circumstances also revised priorities: Doing art became more important. The artists also had more time to discover and try new techniques and mediums, which also helped develop their craft. They felt free to try out different things, to be experimental, to become more (or less) abstract, and to revise their subject matter. These shifts, along with feelings of optimism and renewed motivation, compensated for losses in physical abilities, lessened energy, and reduced sensory capacities. (The latter were directly addressed in another part of the questionnaire and are reported in Chapter 13.)

About half of the reasons given for late-life increases in creativity were related to the self, or personal: improved knowledge, self-acceptance and -understanding, a greater concern for excellence, enhanced motivation, and increased maturity. The remaining reasons, equally divided, pertained to work (new techniques, shifts in subject matter, increased abstraction or the reverse, and more experimentation) and external circumstances (decreased family responsibilities, growing demands from gallery owners, and having more time to work). The most prominent reason for changes with age was increased knowledge, which appreciably influenced four aspects of creativity. Other reasons impacted on one aspect of late-life creativity more than others. Self-acceptance had its greatest influence on the origins of ideas; having enough time to work influenced mainly quantity; and decreased family responsibilities affected the quantity of work much more than its quality. The factors responsible for changes in late-life creativity overlapped to some degree but not completely. Creativity is indeed multidetermined.

The reasons for change, moreover, were rather straightforward, practical, and mundane rather than subtle, profound, or grand. Gains in late-life artistic creativity were often the result of increased information, knowledge, and skills. Little or nothing was said about shifts in the artists' philosophy of life or personality, challenges arising from growing old, or the need to make one's mark as life neared its end. Few wrote about intuition, insight, or the unconscious; allusions to wisdom were rare. In short, the reasons omitted were as interesting as those mentioned.

The absence of dramatic and arcane explanations for gains in creativity with age does not mean that unmentioned factors were unimportant or nonexistent. Omissions reflect the limitations of self-completed questionnaires. Had the artists been interviewed and subject to in-depth probing of the unspoken and unwritten, their replies might well be different. The advantages of a more depthful methodology assumes a great deal, though: Covert and non-conscious thoughts can be evoked by face-to-face queries, by a stranger, no less; and complex responses can be adequately summarized and consensually interpreted, not to mention quantified. A large order, indeed.

All research, whether qualitative or quantitative, poses dilemmas without easy solutions. What if experts, looking at the artists' work, unanimously (!) disagreed with the participants' assertions that it became better with age? I doubt if the artists (or other viewers) would change their minds. There is some justification for taking artists' replies at face value. "Creatives may themselves be the best qualified to judge their solutions and products.... They are the best informed" (Runco, 1995, p. 386).

Critics of the research might pose this objection. "The artists purposefully slanted their replies in order to present themselves in the best possible light. How many would admit that their work (and they) became worse with age? Thus, the artists inflated their creativity, themselves, and their art in order to preserve an idealized self-image." The critics might also question the overwhelmingly sanguine reports of this particular sample. After all, they would point out, a fairly substantial number of nominated artists (45%) did not complete the questionnaire. Like van Gogh, critics might argue, non-participants are tortured and tragic souls whose uncooperativeness reflected their pessimistic view of the course of their late-life creativity.

It is true that self-reports and objective measures can be contradictory. "Older adults commonly report growth in practical abilities over the years, even though their [tested] academic abilities decline" (Sternberg, Wagner, Williams, & Horvath, 1995, p. 913; the authors were referring to a study by Williams, Denney, & Schadler, 1983). Thus, test performance on the cognitive abilities of older adults peaked early in life, yet 76% of the testees believed their ability to think, reason, and solve problems increased. But when the discrepancy between tests and self-reports was pointed out to the older adults they were unfazed. They explained that they had been given "school" problems rather than practical problems from everyday life in which they did excel. Inconsistencies between subjective and objective measures are difficult to resolve.

Not easily answered, too, is the criticism that the artists who completed the questionnaire, although judged creative by their peers, did not include, as far as can be determined at this time, great artists whose works will eventually be enshrined in museums. Had such potential luminaries been somehow recruited, critics might contend, they would have given different and more negative accounts of their late-life creativity.

The charge of a biased sample is difficult to rebut, since it is impossible to say which artists will become renowned in the future. Notable artists of the past, like Rembrandt, were relatively unrecognized in their lifetimes; others died in obscurity only to have their reputations resurrected two or three generations later. The history of art is replete with examples of artists for whom a considerable amount of time had to pass before their status was established, lost, or regained. It may take fifty or more years after artists' deaths for reliable judgments of late-life creativity.

Countering accusations of biased self-reports and limited sampling is the consistency of the results. Improvements with age were reported by three age groups, for both men and women, and in response to four questions about different aspects of creativity. Does it seem reasonable for up to 90 artists, generally known for their individualism and nonconformity, if not cantankerousness and stubbornness, to give uniformly self-serving reports? This question might be answered by longitudinal research: What would 60-year-olds report about their creativity at 80 (assuming they lived that long and were available for retesting)? A limited answer is available from the retrospective ratings of 80- and 70-year-olds to their creativity at earlier ages. They were uniformly positive.

Self-reports certainly have to be cautiously weighed and the future status of the participating artists must be held in abeyance. Left for future research, too, are other questions. Unknown, for example, is whether improvements with age also occurred in ordinary daily activities and in the artists' relationships with others. Were there offsetting losses in social areas? Subsequent research, therefore, should include the reactions of spouse, family, friends, and colleagues to the gains reported by artists in their work. Increased creativity may be a mixed blessing.

Some reservations about the research are therefore in order, and alternative interpretations of the findings are possible. Conclusions about the late-life development of artistic creativity must be tentative rather than definitive, but greater certainty will come with subsequent research. For the moment, though, it seems reasonable to accept this fairly well-established fact: Recognized 60–80-year-old creative artists emphatically asserted, in overwhelmingly optimistic terms, an encouraging view of growing old.

The positive relationship between aging and art could be attributed to living in favorable times (Chapters 7 and 8). Contemporary artists have better health care, more educational opportunities, and greater financial security than their predecessors. Living in a highly developed and advanced society like the United States in the 21st century enabled artists to work longer, harder, and better than their forebears. This is not to argue, though, that good social and economic conditions necessarily result in a surplus of great artists who produce a cornucopia of creative works (Simonton, 1976a, 1976b). Most will agree that today's artists, despite improved material circumstances compared to the past, have a harder time pursuing their careers and earning a living than members of many other professions. Some creative individuals, furthermore, may need troubling times, and indeed, have to "suffer for their art," in order to do great work (Jamison, 1994; *cf.* Lindauer, 1994).

The consistency of the results, noted earlier, bolsters the validity of the findings. So, too, does the congruency between contemporary artists' replies to a questionnaire and productivity counts for historical artists, which also increased with age. These facts are also congruent with the late ages at which masterpieces of art were produced by major artists of the past (Chapters 7 and 8). Growing old for artists at different times and places did not mean a loss of artistic creativity. Thus, multiple lines of evidence, based on several samples of artists, both historical and contemporary, as well as studies of nonartists yet to be reported (Chapters 15 and 16), along with methods of research that include productivity counts and questionnaires, and subsequently, surveys and experiments, point to a positive relationship between aging and creativity.

It is not the 30s, or even the 40s or 50s, when creativity flowers and wanes. For aging contemporary artists, personal testimony on several aspects of their late-life creativity, based also on quantitative ratings, indicated steady increases in the quality and quantity of art, improvements in the way they approached their work, and the emergence of new ideas.

THE FUTURE

The facts about historical and contemporary artists counter a pervasively negative outlook on late-life creativity. Decline with age is not the only option. Positive results for painters prompts a closer look at other kinds of aging artists. Would old composers, novelists, dancers, and filmmakers also give optimistic accounts of their late-life creativity? Investigations of different old artists would also expand our knowledge about the elderly in general, thereby liberalizing current views of aging.

Unfortunately, elderly artists are rarely studied, at least with empirical tools sensitive to sampling, methodology, and quantification. Instead, informal interviews, anecdotes, autobiographies, and other qualitative sources are the usual sorts of information we have about aging artists, along with critical analyses of individual works and

artists by scholarly experts. The exclusion of old artists in formal
research on late-life creativity is a serious omission (Chapter 17). Aging
artists who maintain an active and productive career demonstrate
robust abilities.

The continued creativity of aging painters raises doubts about the
early peaks in creativity and declining productivity found for other professions (Lehman, 1953) and the low creativity test scores of the elderly
in general (Chapter 3). Highly competent aging professionals outside of
the arts, such as old scientists and elderly scholars, are underrepresented in research on old age. If they were more often included, they
might provide more positive information about aging, as artists did.
Aging scientists may increase their publication rate, improve the quality of their research, redirect their methods, change their topics of interest, join collaborative efforts, shift from the lab to writing insightful
critical essays and thoughtful reviews, become administrators of grand
projects, and head important agencies. Similarly, old scholars may
write more incisively, eloquently, and broadly about new subjects and
in uniquely personal and identifiable styles.

Open to further inquiry, too, are late-life changes in aging non-
professionals. When they retire, some turn to painting and other forms
of artistic and creative expression, including wood-working and quilt-
ing, while others start new careers or take on new challenges as volunteers. Studies of contemporary painters suggest that creative work late
in life is associated with increased self-knowledge, improved relationships with others, and an enhanced world view, among other benefits.
Similar consequences might be harvested by elderly nonartists through
an engagement with creative endeavors in the arts and other outlets for
self-expression.

Studies of aging artists therefore open up several new avenues of
research and discussion, lend additional credibility to the sustained
nature of late-life cognition and other abilities, and demonstrate still
another avenue for successful aging. The glowing reports of old artists
involved in a lifetime of creative work demonstrated that excellence
in old age is possible, continued learning through the 80s can occur,
and changes with age can be for the better. Declines in old age are not
inevitable and stability is not the only option; gains, too, can be
expected. For artists who continued to be creative beyond their 60s,
old age had tangible benefits, and these should hold for nonartists as
well. Higher-order abilities required to do art, like solving aesthetic,
practical, and personal dilemmas, and calling upon imagination and
imagery to solve problems, can be sustained, refreshed, and improved
with age. These competencies are not limited to art, of course, but

apply to all sorts of tasks and goals with similar complexities, requirements, and goals. The continuity and growth that mark late-life creativity, demonstrated in this Part, foster a positive perspective on growing old.

Three possible outcomes of late-life creativity were empirically verified: decline, continuity, and growth. These diverse (and opposing) trajectories remind us of the multifaceted nature of creativity. This expansion of possibilities should motivate scholars, art historians, psychologists, gerontologists, cognitive scientists, administrators, and writers to pay more attention to late-life artistic creativity, late art, and old artists, and the positive consequences of creative endeavors when growing old (Chapter 5).

Late-life artistic creativity can take at least one other path. Aging artists might pursue an entirely new way to express their creativity. They might radically modify or expand previous techniques or develop novel ones, experiment with different materials, or invent a new medium or unusual venue for their art. The questionnaire did ask about changes in the aging artists' approaches to art as they aged, and while the general answer emphasized improvements, not indicated was whether they were consistent with previous ways of working, modified or revised earlier modes, naturally evolved, or took sharp and unpredicted pathways. A drastic change in the expression of artistic creativity with age is called the old-age style, a shift in late-life creativity that is more complicated than the other kinds of changes reviewed so far. The old-age style might reflect the decline of creativity, or alternately, wrench it into a new direction, which implies both continuity and improvement. The nature of the old-age style is covered next, in Part IV.

The Old-Age Style

It is too late! Ah, nothing is too late
Till the tired heart shall cease to palpitate.
Cata learned Greek at eighty; Sophicles
Wrote his grand Oedipus, and Simonides
Bore off the prize for verse from his compeers,
When each had numbered more than four-score years ....
Chaucer, at Woodstock with the nightingales,
At sixty wrote the Canterbury Tales;
Goethe at Weimer, toiling to the last,
Completed Faust when eighty years were past.
These are indeed exceptions; but they show
How far the full-stream of our youth may flow
Into the arctic regions of our lives ....
For age is opportunity, no less
Than Youth itself, though in another dress,
And as the evening twilight fades way
The sky is filled with stars, invisible by day

Longfellow, *Morituri Salutamus*

CHAPTER 11

The Old-Age Style
Introduced: The Issues

With years a richer life begins,
  The spirit mellows;
Ripe age gives tone to violins,
  Wine, and good fellows.
        John Townsend Trowbridge,
        Three Worlds

Creativity of the artistic kind extends well beyond youth and into old age (Part III; Lindauer, 1999). But is there a *special* kind of artistic creativity in old age? Stated more generally: Can creativity change drastically or be different late in life? Art historians have answered with a qualified "yes," and point to an old-age style among some artists, which like wisdom, emerges late in life. Unexpected shifts occurred in the techniques, composition, subject matter, and affective tone in the later works, for example, of Cézanne, Goya, Michelangelo, Rembrandt, and Titian.

Something akin to an old-age style might exist in other professions, too. Aging scientists "purposely place themselves in the position of becoming novices again every five or ten years [and] become mentally young by starting over again [with] the courage to be ignorant again" (Root-Bernstein, Bernstein, & Garnier, 1993, p. 341). An old-age style might also exist among scholars (Birren & Deutchman, 1991; see also Munnichs, 1990) and for the elderly generally it might be reflected in the attitudes they take toward their own creative efforts (Pruyser, 1987) and the kinds of art they enjoy (Butler, 1973). The old-age style might also account for late-bloomers' sudden and unexpected creative achievements (Goldman, 1991) and aging "eccentrics." The phenomenon could also be the basis for the "vital involvement" and "creative living" promoted by Erikson and his coworkers (Erikson, Erikson, & Kivnik, 1986) that accounts for significant late-life changes (see especially pp. 324 and 363–364). An old-age style, too, might underlie the major occupational changes in the later years of Terman's gifted children (Holahan, Sears, & Cronbach, 1995; see also Chapter 5).

About 3% radically changed their occupations after retirement:
A newspaper reporter became a lawyer, a professor turned to banking,
a farmer developed into a professional bridge player, a manufacturer
took up poetry and then painting, an investment analyst entered the
professorate, and a sports cartoonist switched to teaching and then to
painting.

The old-age style should therefore be of considerable interest to
gerontologists, developmental psychologists, and other professionals
interested in extreme forms of successful aging (Chapter 1). Marked
changes in creativity, if validated empirically, demonstrate that abilities of a high order can be initiated, reinvigorated, and redirected
in old age. The old-age style also counters the decline position, which
sees the later years as a time of losses, as well as modifies the continuity argument; the development of late-life creativity is not a straight
line predictable from earlier forms. The study of the old-age style,
moreover, is another opportunity for artists, art historians, literary
critics, scholars in the humanities, and others in the arts to join with
psychologists, gerontologists, and scientists in investigating aging.

The old-age style is discussed mainly in the arts, including sculpture, music, and film (Cohen-Shalev & Rapoport, 1990), and extensively
in literature, for example, in the late works of Tolstoy, Goethe, and
Cervantes (Brinkmann, 1925; see especially p. 33) and Ibsen (Cohen-
Shalev, 1992). Late poems, for instance, are "characterized by a calm,
meditative tone, and an increasingly personal voice [and] tend to be
angry or cynical [or] influenced by events in the life of the writer"
(Adams-Price, 1998a, p. 278). In comparing the old-age style of writers
and painters, Clark (1972) concluded that only the latter became better
with age as a reaction to despair, misery, pessimism, and rejection
(painters also declined less than writers).

The old-age style, though, centers mainly on painters. Discussed
are a number of related issues (Lindauer, 1992, 1993a, 1993b, 1998a):
(1) A definition and (2) a description of works done in the old-age style;
(3) the identification of aging artists and works who changed this
way; (4) the incidence with which these changes occurred; (5) whether
works done in the old-age style are improvements over youthful
efforts; (6) the difference between historical and contemporary forms of
the old-age style, if any; (7) an explanation of why some artists have an
old-age style and others do not; and finally, (8) the presence of an old-
age style in different areas of the arts, outside of the arts, and distinctions between the various forms it takes, if any. These issues are
discussed in this and the following chapters of Part IV, along with
research on these matters.

DEFINING AND DESCRIBING THE OLD-AGE STYLE

The old-age style, or "Altersstil," a term coined by Brinkmann (1925), has received extensive coverage in art history texts (Munsterberg, 1983), special journal issues (Youthful works by great artists, 1962; Benn, 1955; Berlind, 1994; Brinkmann, 1984; Hood, 1977–78; Rosand, 1987; Rosenblum, 1977–78; Rosenthal, 1954, 1968; Spencer, 1962; Talley, 1990; Tietze, 1944–46), and exhibitions (Allen Memorial, 1962; Feldman, 1992; Hufford, 1987; *Man and His Years*, 1954; Rosand, 1987; Rosenthal, 1968). Despite this attention for more than 75 years, uncertainties abound in its definition, description, and other matters.

Art historians generally agree that works with an old-age style, compared to youthful efforts, are characterized by changes that are "sharp," "marked," and "drastic," but the meanings of these terms are not spelled out, differentiated from one another, or distinguished from other labels, like "developed," "matured," and "evolved." For example, the seminal work *Man and His Years* (1954) noted that some artists "changed," "changed drastically," "developed a distinctive style," "added a new facet," "stayed the same in some ways and differed radically in others," and "broadened or concentrated earlier tendencies." These kinds of change are not differentiated.

There is some vagueness, too, in defining "old." Art historians consider works done after age 60 as about the time when an old-age style should emerge, if it is to do so. But 60 is a more appropriate age for earlier times when the span of life was much shorter than today's. Even 60 is not consistently used. A major source of late art, *Man and His Years* (1954), includes 11 artists (16% of the 70 covered) who died when they were around 40 years old (e.g., Bellows, van Gogh, and Pollock). The late works of these short-lived artists were included because they shared certain properties with the old-age style of long-lived artists. The label "late-life style" is occasionally substituted for the old-age style when applied to short-lived artists (not to be confused with "late art," which refers to the historical development of different types of art, as in abstract art being a later form of art than representational art). Artists' depictions of "the stages of life" add to the ambiguities surrounding old age when they segment the course of life into three to twelve parts, suggesting that the "artistic clock" for aging is variable.

More difficult to define than "change" and "old" is the term "style," which refers to the essential, underlying, and uniquely distinguishing features of an artist's work (Schapiro, 1953; see also Chapter 15).

Style depends on an intuitive grasp of complex and implicit features that characterize multiple works produced over an artist's lifetime, a highly difficult and subjective task. In the case of the old-age style, the difficulties are doubled, so to speak, for earlier and later styles have to be distinguished. Statements about style may be apparent when pointing to examples, but are not easily verbalized.

Given so formidable a task, art historians understandably prefer to concentrate on one or a few works or a handful of artists. Consequently, different artists and works are selected to illustrate the old-age style. Not surprisingly, scholars disagree on their choices, hold incompatible views, reach conflicting conclusions, and come to different conclusions. Some examples follow.

Older and younger works, to paraphrase Brinkmann (1925), are distinguished because they

> differ in their diversity of their forms, colors, and subjects. For example, the colors of the older Titian are gray, monotonous, and muted, and their forms blend into shadows; Tintoretto's themes are tranquil; El Greco interpreted forms in new ways; and Renoir changed his backgrounds. Younger and older artists also differ in their introvertedness, that is, their thoughts, as expressed through their work, turn inward to self or outward, respectively.

A different emphasis is given in this paraphrasing of Gantner's (1965) position:

> Works done in the old-age style are dynamic and mutable, that is, not static or constant; they look ahead, rather than back; they involve newer and less comprehensible forms, and hence are unpredictable, freer, and more expressive; they are also fragmentary, without form, sketchy, and more intense. There is a disregard for clearly articulate forms.

Other authorities focus on one particular quality. The amount of detail, for example, is reduced in works with an old-age style. Alternately, the brushstroke is broader and freer than youthful efforts, signs of "the assurance and freedom of... elderly artists... more concerned with total effect than specific details" (Gedo, 1984, pp. 334–335). Another expert sees the brushstroke, among other features, as indicative of the artist's state of health which in turn leads to an old-age style. The "special qualities of brushwork and colors and pictorial structure [reflect] the old artist's physical deterioration" (Rosand, 1987, p. 91).

Art historians amplify these descriptions, add new elements, or rephrase them. In *Man and His Years* (1954), for example, Rosenthal maintained that older works were "pervaded by a spirit of inner serenity... a freedom of conception and execution which goes far beyond the daring innovations of their youth [a d]irectness of visual statement and the absence of self-consciousness [that reflect individuals who are] rarely driven by ambition for public recognition" (p. 5). Breskin (1954) emphasized the artist's freedom. The "late works [are] often freer and bolder than [the artist's] younger years. They throw off restraints and timidities and [there is the] full satisfaction in uninhibited self-expression" (p. 5). Some descriptions are extensive enough to be summarized as a list: brilliant, contented, conventional, dignified, dramatic, freer, fresh, harmonious, insightful, peaceful, positive, self-possessed, serene, simpler, spirited, tranquil, and wise (Munsterberg, 1983). Another list includes "radical honesty" and "an abandoning of the relations between people," along with inwardness, gentleness, melancholia, mellowness, reserve, ruthlessness, spirituality, toleration, wariness, weariness, and weightlessness (Benn, 1955).

The holistic quality of the old-age style was emphasized by Arnheim (1986a, 1990). There is "a lack of differentiation, a non-separation between things, self, others, persons, and events; it is something like Freud's 'Oceanic feeling,' a turning away from specifics to generalities, to an overall experience" (personal communication, September, 1989). Holistic characteristics, too, describe the way artists handle human figures and treat formal qualities like light; it is not as focused as it was in youthful works but pervades the whole scene; it is vaguer, more homogenous, subdued, and less distinct. Arnheim illustrated undifferentiated qualities in the older works of Michelangelo, Rembrandt, Tintoretto, and Titian, where there were shifts from the dramatic, active, contrasting, distinctive, physical, and "busy" strengths of the younger artist to the more tranquil, inner directed, and darker visions of the older artist. There is, Arnheim amplified, a movement from broad perceptions, understanding, and generalities, with little differentiation of the self and others, to a more detailed, looser, diffuse, and less literal emphasis. (For the holistic perceptions of long-lived experts in any field, see Cooke, 1990.)

Shifts from youthful preoccupations also describe the old-age style of composers and filmmakers. "The young artist shifts from a concern with human beings and emotional reactions to the natural world, so that in old age there is a more abstract concern with appearances" (Cohen-Shalev, 1989b). There are more formal or textual changes. Thus, surface coherence

disintegrates so that there are gaps, sudden transitions, fragments, and a confusion of conventional genres. The older art therefore expresses discontinuities and dedifferentiation; it is inward directed and introspective rather than action-oriented; there is a withdrawal to the inner world of experience. The art is passive and ambiguous as the artist reminisces on the meaning of art for one's life and for humanity at large. (p. 21)

Questions of definition and description are further complicated by authorities in various fields who write about the old-age style. Within the visual arts alone, the old-age style is applied to painters, architects, and sculptors. In other areas of the arts it is attributed to composers, authors, and filmmakers. Each perspective adds new layers of definition and description. Some accounts, furthermore, are infused with a psychoanalytic orientation; and psychological perspectives are either empirical or clinical. To these multiple viewpoints add historical, sociological, and economic changes between artists' youth and old age that affect late-life art.

A number of leading authorities with different viewpoints have therefore proposed a diversity of definitions and descriptors of the old-age style. Some are narrow, focusing primarily on technical aspects, such as the brushstroke or the amount of detail. Others emphasize the broad characteristics of the work as a whole, such as its undifferentiated forms and a holistic treatment of light. Still others shift from a focus on the work itself to the artist's personality, and interpret the old-age style as a reflection of joy or pessimism, health or deterioration. A few turn from the characteristics of the artist and the work to the viewers' reactions, which includes puzzlement and surprise. Nearly 100 different ways of defining and describing the old-age style infuse the literature (Chapter 13), making it easy to understand the lack of consensus on its explanation and other matters, such as identifying artists with an old-age style.

IDENTIFYING ARTISTS WITH AN OLD-AGE STYLE

Disagreements over which artists have an old-age style are the inevitable result of fuzzy definitions and bloated descriptions. Consider the artists discussed in two major references to late art. Munsterberg (1983) and Rosenthal (1968) refer to 90 and 70 long-lived artists, respectively, assign an old-age style to 35 (22%), but agree on only 13 (37%): eight had an old-age style and five did not. A third major source on late art, *Man and His Years* (1954), refers to 16 older artists in the Munsterberg text and 17 in the Rosenthal collection. For the 15 artists discussed in the three sources, agreement occurs for just four: Cézanne, Goya, Marin, and Titian had an old-age style. The other 11 artists (Chardin, Corot, Degas, El Greco, Hartley, Homer, Manet, Michelangelo, Pissaro, Renoir, and Rubens) were inconsistently judged. Diverse opinions are also found *within* the Rosenthal (1968) collection, between the editor and the specialists who wrote about Copley, El Greco, Grueze, Kandinsky, Leger, Matisse, Mondrian, Reynolds, Sargent, and Tobey. Similar disparities between the experts characterize other works about artists with an old-age style (Berg & Gadow, 1978; Boggs, 1977–78; Edel, 1977–78, 1979; Gantner, 1965). Thus, depending on the authority consulted, different artists are identified as having an old age-style; disagreements are the rule.

The problem could be reduced if there was more agreement on defining and describing the old-age style. But this goal is difficult to achieve. Experts do not explicitly state the criteria they use to select certain artists *and* for excluding others. The Munsterberg (1983) text is illustrative. Eight groups of artists displayed different kinds of changes with aging, of which one set exemplified the old-age style. No indication was given, though, on how the choices were made for each of these eight types of changes with age, how the groups were constituted, why they were labeled one way and not another, or the reason they were limited to eight.

Discussions of artists with an old-age style can also be faulted for failing to contrast artists with and without an old-age style, that is, those whose style did not change with age. To ignore the absence of an old-age style (or any other kind of late-life development) is a serious omission. Some baseline (a so-called "control group") is needed in order to make comparisons with artists who did change. Without a contrasting group of artists, the crucial traits of artists with an old-age style cannot be ascertained, since seemingly critical traits cannot be distinguished from those common to aging artists in general. Many old artists could have, for example, a rough brushstroke or a pessimistic attitude toward life, but only if these characteristics were absent in artists without an old-age style could they be considered key distinctions.

Specifying artists without an old-age style also makes it possible to estimate the relative size of artists with an old-age style, thereby giving the phenomenon some magnitude. Is the number low, modest, or substantial? Let us say that 20 artists have an old-age style. This is a minor number when weighed against the hundreds of old artists who lived and worked at the same time and place but 20 is significant when compared against 50.

A numbing number of practical difficulties also impedes clarification of the old-age style. There are hundreds of artists and thousands of works, and many artists worked in several mediums. Which artists and works should be analyzed in searching for an old-age style? Examples of artists' early and late works, moreover, are often available only as black and white reproductions, as prints or on slides, where color distinctions and other important details are lost; and even color reproductions are often of questionable fidelity to the original. Originals may be inaccessible in private collections or distant museums, or difficult to locate, and once found, they may have deteriorated in quality. Only a relatively few reproductions of early and late works, small in size and impoverished in color and details, can therefore be inspected for signs of an old-age style.

With so many conceptual and pragmatic problems, no wonder experts differ in their definition of the old-age style, the artists who have it, the description of such works, and even on the basic question of whether there is an old-age style.

IS THERE AN OLD-AGE STYLE?

Proponents of the decline position (Chapter 3) do not expect an old-age style, or much of one, since they see creativity as peaking and ending rather early in life, around age 30 or 40 at the latest, well before old age. Productivity counts of the old-age style are problematic, too, since this type of change occurs rarely among a few (the old), and in any event, qualitative shifts are difficult to discern when counting large masses of data. For similar reasons, signs of the old-age style are unlikely to translate into high scores on creativity tests or to emerge from questionnaires and interviews (assuming it could be clearly defined and accurately described).

Nonetheless, few art historians doubt the universality of the old-age style (Soussloff, 1987). But there are some who do (Levy & Langer, 1999, pp. 50–51). Feinstein (1988, pp. 3–4) claimed there were no special qualities of the art of old age that deserved a distinctive cachet. The old-age style, she insisted, is only a label that tells us more about "critical and aesthetic trends, about the impact of social history," than the creative process. However, some doubts are hinted at when she adds, "This is not to say that all works by artists who continue to create in old age, however impressive, bear the hallmarks of the old age style while, conversely, the last works of some artists who never reached old age do." She also accepted one class of exceptions, "artists who continue to work despite deteriorating vision and loss of physical dexterity and are thereby forced into a personal old age style."

A skeptical but equivocal attitude was also taken by Foote (1994, pp. D1 and D4) in reporting on a show of paintings, sculpture, and drawings entitled "Still Working," at the Corcoran Gallery of Art in Washington, DC by "underknown artists of age [in their 70s]." On the one hand, Foote asserted that these works "never focus on the theme of age," but instead, go in various directions, which "debunks the theory of 'old-age style.'" On the other hand, Foote admits that the late works revealed several new features, including a "freedom and independence that seemed to come with age. Some are doing the best work of their lives. They have all given up on being right or wrong.... They have stopped asking why, and they really don't care that much about what other people think. [There is] a lack of concern about outside opinion."

Interviews of artists are also mixed. No evidence of an old-age style was found in Berman's (1983) interviews of 70–80-year-old artists. Instead, changes in style were gradual, natural, and evolutionary, and a function of accumulated experiences rather than old age, as such. Berlind (1994, pp. 7 and 21), too, is convinced that no one pattern fits artists who he interviewed. "[L]ife continues in all its unpredictable diversity and eccentricity." Further, age, per se, does not define an artist, just as gender, ethnicity, or sexual orientation do not. "The interviewees did not, for the most part, claim [an] old age [style] for themselves.... In fact, none of the artists acknowledges any special agenda relating to their age. Most say that they simply continue to do what they have always done [revealing a] unanimous dissent from the notion of *Altersstil* as a general proposition." Yet, a subtle feature of the old-age style may be at work when Berlind complained, "We consistently underrate the work of older artists" because of an overemphasis on their earlier works, when they first matured or came to the attention of the art world.

In contrast, Blythe (1979) reported that some of the aging artists he interviewed admitted to the presence of old-age style. Indications of an old-age style were also reported by nonartists who were sent an informal questionnaire by Hall (1922). One item implicitly referred to the old-age style: "Did you experience an 'Indian summer' of renewed energy before the winter of age began to set in?" (pp. 347–350). Some respondents reported such a change.

Kastenbaum (1992) is cautious about the old-age style, not so much its existence as its explanation as a function of age. "Age per se may have relatively little to do with the emergence of a late style." He suggested other circumstances that might lead to late-life changes, but warned "against hasty conclusions and undisciplined generalizations" (p. 301). A mixed view is also held by Simonton (1989, 1990b, 1994). He saw some evidence for an old-age style in the secondary peaks of late-life creative productivity, although he emphasized that these were not as high as the younger years. Simonton also reported on composers who changed their style when they sensed a depletion of creative potential as they grew older and approached death, to which they responded by creating a lasting memorial, or "swan-song." Simonton concluded that changes with age can affect artistic expression positively, leading to a greater openness to and expression of new ideas, but there are negative consequences, too. Late-life art could deteriorate because of losses in health, physical and mental decline, and decreased energy.

The old-age style may be denied by some because it is easy to miss; the phenomenon is rare. It is also clearer among historical artists, when representational art predominated, but is difficult to discern in modern art. Monumental changes have occurred in art, especially the movement away from realism to abstraction. Many of today's artists, too, are disinterested in drawing accurately and in displaying technical virtuosity. The old-age style may not therefore show up among contemporary artists, or at least not until they reach a very old age. Contemporary artists live longer than their predecessors, so they would have to live to a very old age, say 80 and not 60, as in former times, before an old-age style might emerge. But fewer artists are likely to be alive or working at 80, even under current high standards of living, thereby reducing even further the likelihood of an event that is already rare. If and when it did emerge, it would look quite different from the kinds of changes noted among historical artists. Thus, critics are likely to miss contemporary versions of the old-age style since they are looking for discontinuities in traditional ways, such as brushstroke or detail, features developed by early masters but which are no longer relevant to modern artists.

The composition of audiences and patrons, and their attitudes toward old artists, late art, and the old-age style are also different from earlier observers, perhaps because they, too, live longer, among other reasons. These differences could affect viewers' discernment and evaluation of an old-age style. Complicating matters, too, today's artists are in greater communication than their earlier counterparts with fellow artists throughout the world, through international shows, the availability of a large number of relatively inexpensive books, and electronic means. Contemporary artists also have additional benefits not found in the past: improved medical care (Medicare), greater financial resources (Social Security), more sources of patronage (government grants), increased outlets for their work (outdoor fairs, open houses, "clothes line shows," and sidewalk exhibitions), and additional ways of supplementing their income (teaching, working in senior centers, heading workshops, running programs for the mentally challenged, and doing art therapy). Thus, today's aging artists have greater opportunities and face different pressures than their historical predecessors, and these probably affect the incidence as well as the forms taken by the old-age style. Moreover, modern artists, according to Rosenblum (1977–78), "create their finest work in youth and then [become] self-imitative or... irrelevant to contemporary developments" (p. 68).

It is therefore extremely difficult to identify features of works done in the old-age style among contemporary artists and probably impossible to compare them to those done in the past in order to determine if works of this sort exist today. For these reasons Gedo (1984) has urged an updated version of the old-age style to reflect changing times. Others are less hopeful about a revision. Observers of the art scene find it sufficiently difficult to judge the merits of current efforts, let alone the existence of an old-age style, which is a rather specialized and rare phenomenon (Cohen-Shalev, 1989b). Drastic changes in today's aging artists' style, if it occurs, would be received, as they were in the past, with puzzlement, as unexpected and strange compared to youthful efforts. It would therefore be judged harshly and not honored. The likelihood of unusual works' premature rejection led Brinkmann (1925) to warn about overly critical judgments of late-life works (see especially p. 58). Discrepancies from youthful accomplishments, rather than seen as signs of a reinvigorated art, can easily be interpreted as the result of some breakdown or deterioration, and indicative of illness, fatigue, senility, or eccentricity. Misinterpretations of art in the past have often led to its rejection, and we can assume this is still true.

Consider the 16th century Vassari (Lavin, 1967), one of the most highly respected figures in art history. His *Lives of the Most Excellent Painters, Sculptors and Architects* is a highly influential work in the history of art. In it, Vassari expressed the wish that Titian, one of the greatest artists of the Western world, should have stopped painting years earlier because his early works were the best and the new style of his later works detracted from his reputation. As it turned out, the late paintings of Titian are considered prototypical examples of the old-age style and are universally judged as among the finest he or any other artist ever produced (personal communication, August, 1997, from an anonymous reviewer of one of my articles).

Judgments about today's art, whether it has an old-age style or some other feature, are subject to a considerable amount of error. Reliable appraisals await the informed opinions of later generations of historians long after an artist's death, when more balanced and tempered evaluations are possible. The future (re)appraisal of late works may revise evaluations of works done in an old-age style that were rejected in the artists' lifetime. Like the reputation of any kind of art or style, and the status of an artist from earlier eras, the final judgment of an old-age style depends on the passage of time. This is not to say, though, that current research cannot make some headway on definitional, descriptive, and other issues (see Chapters 13–16).

Despite a few reservations from some art historians, most believe there is an old-age style. "One can change centuries but the contrasting periods [of young and old art] repeat themselves" (Brinkmann, 1925, p. 19). This certainty was echoed by Rosenthal (1968). There are "certain related tendencies that may be detected in the late works of great masters who retained their creativity in old age" (p. 12). The noted art historians Clark and Blount (reported in Kemp, 1987) were confident that the old-age style revealed personality changes that inevitably occurred with old age. Similarly confident, Kemp (1987) maintained that detachment from work, pessimism, and anger characterized the work of "almost every great artist who has lived beyond the age of 65 or 70" (p. 94). The universality of the old-age style was affirmed by Tietze (1944–46, p. 177). "The simple fact of being young or old produces a certain kind of art, due to certain age-related mental and artistic attitudes," and these are expressed by "aloofness, a lack of interest in praise, a despair of acceptance by the public." The old-age style, he continued, represents "an Indian summer, an apotheosis, a reaction to [the] public and contemporaries." These characteristics, Tietze insisted, were not due to "physical weariness or mental decline."

The existence of the old-age style is also affirmed by many special exhibitions. Particularly interesting was a 1992 show at the National Gallery of Art in Washington, DC entitled "I am still learning: Late works by masters" (Feldman, 1992). The 17 artists represented in the exhibit were Bellini, Calder, Cézanne, Corot, Degas, El Greco, Hals, Homer, Ingres, Magnasco, Matisse, Monet, Pissarro, Rembrandt, Tiepolo, Titian, and Turner. These works demonstrated that "old age can be a meaningful culmination … in the lives of artists in all fields who continue to practice their art with no loss of talent or inspiration, *often breaking new ground late in life*" (emphasis added). The persistence of creativity into old age, and in changed forms, was explained in motivational terms. "What allows many masters to maintain their creativity undiminished or even heightened in their later years? Can it be a consuming sense of purpose?" The answer: It is the result of "a mission, an impatience to sum up life's experiences coupled with a strong confidence in instinct that has been tempered by trial and error" (Feldman, 1992, pp. 3–4). Inexplicably, the late works of Goya were not included in the show, although he is generally known for his old-age style. The omission is even more surprising in light of a statement he made about the continuation of learning, when he was about 80, that was the title of the show.

A rare empirical study of the existence of the old-age style with over 200 participants, aged 3–80, including both participants untrained in art and professional artists, instructed them to draw human figures. Certain variations in the drawings with age were interpreted as indicators of an old-age style. "[C]hanges in line use during the later years of adult life [exhibited] an increasing looseness of definition and decreasing differentiation between objects" (Spaniol, 1992, p. 93).

Granting there is an old-age style, is it good or bad art, or better than youthful art? As on previous questions, opinions are mixed.

THE QUALITY OF WORKS DONE IN THE OLD-AGE STYLE

The old-age style may reflect aging artists' debilities or strengths, a weakened or reinvigorated creativity, an unhealthy or healthy reaction to aging's losses. Most experts, though, evaluate the change positively. Arnheim (1986a, 1990) insisted that the old-age style was a significant, meaningful, mature, and valuable form of art. Breskin (1954) went further and asserted that the old-age style was proof "of the…enhancing qualities of advancing years" (p. 5). A positive stance was also taken by Berg and Gadow (1978):

> [G]rowing old is an enriching, deepening, experience; it is distinctly not degenerate or [an attempt] to remain young or "nearly normal." It is a time of slowing down, of rejecting the physical, material world as ideal, of turning in, distilling experiences for their essence, of becoming clearer in some cases, or in some cases less certain of knowing it all, and finally, of determining the personal significance of death. (pp. 88–89, 92)

Edel (1977–78), too, interpreted the old-age style as a profound consequence of aging. Artists who live long are "a finished oak, standing strong…most questions have been answered…the pain [has] been surmounted. The…suffering has ended in triumph and fulfillment, even if sometimes in penury and want" (p. 54). The artist Braque (quoted in Greenberg, 1987, p. 28 and Rosenthal, 1954, p. 13), summed up the beneficial effects of old age on art: "With age, art and life become one." The vigor of late-life art was attested to by Picasso, who was quoted to have said late in life (he lived to 92), "Now that I have achieved a great age, I might just as well be 20." Glowing terms also characterized Cohen-Shalev's (1989a) comments about the older operas of Verdi.

> There is compression and urgency, rather than the spaciousness of his youth; there is a flowing line and biting ring, rather than straightforward

> lyricism. Verdi retreats from his earlier psychological realism in favor
> of sketchy, abstract portrayals; he displays an overall impatience with
> established operatic conventions and a symphonic (linear or melodic)
> conception. (p. 11)

Negative views of the quality of art in old age, however, are also voiced. The old-age style is a trade-off with wisdom (Simonton, 1990a, p. 321), with the former decreasing as the latter increases. More pessimistic are accounts of the old-age style that emphasize its source in physical and sensory shortcoming, cognitive handicaps, and personal difficulties in coping with growing old: artists cannot do what they once did and therefore change in ways that observers interpret as the old-age style. Drawing upon his own experiences as a poet and informal observations of and reading about writers and artists, both historical and contemporary, Benn (1955) took a rather bitter, fatalistic, and melancholic view of the consequences of aging. These negative judgments describe the frustrations of aging that afflicted the art of Degas who

> experienced the humiliation and irritations of a long and debilitating senes-
> cence...; he had no family for emotional support, and was often ill.
> Consequently, his work is austere, harsh, and brutal...his dancers express
> weariness, apathy [and] reject the expectancy and artfulness of his earlier
> years [reflecting] the circumstances of his old age...pettiness...loneliness,
> lack of fame...deteriorating eyesight. [Thus, Degas is reported to have said:]
> Count me out "Read Job and Jeremiah and keep going". (Boggs, 1977–78,
> p. 60)

Dour accounts of old age by artists, though, should not be confused with posterity's evaluation of the quality of their late works. Great works of art can be independent of artists' feelings about them. Further, masterpieces can transcend the weakened conditions of aging under which they were produced. The aging and ailing Matisse compensated for failing strength and vision by cutting out innovative collages from large pieces of cardboard. DeKooning, a victim of Alzheimer's disease, nevertheless continued to produce extraordinary works.

EXPLANATIONS OF THE OLD-AGE STYLE

There are almost as many explanations of the old-age style as there are definitions and descriptions, which is understandable, given the variety of artists and works studied. One of the most provocative is by the analyst Jung (quoted in Munsterberg, 1983): the aging artist "turns inward to illuminate one's self rather than the external world for others" (p. 125). What is disclosed therefore becomes the source for the old-age style. Turning inward may also occur because of personal crises that are eventually translated into an old-age style. The aging Michelangelo, Gedo (1984) held, experienced conflicts over religion, while Picasso's social-interpersonal failures led to late-life changes. "In order to enjoy an optimal final phase," Gedo argued,

> an aged artist must be able to identify with some larger force outside of himself. This identification [or] ecstatic merger permits and encourages the elderly genius to view himself and creations [critically with] self-detachment. [Monet and Matisse were able to do this but Picasso] lacking this type of objectivity, failed to achieve a comparable inventiveness. (1984, pp. 358–359)

Personal anguish among literary figures, especially "depression, bereavement, or worries about aging and death," according to Wyatt-Brown (1993, pp. 7–8), are also the cause of an old-age style. Fortunately, "artists are more able than most of us to voice their deepest fears." Some may not, though. As the aging Hemingway became more aware of unacceptable feelings of androgyny, feminine aspects were denied by acting as masculine as possible, but the conflict remained unresolved and became more overt in his later works (Guttmann, 1987). Similarly, the fragmentation and formlessness of the characters in Ingmar Bergman's later films reflected the aging director's own late-life dilemmas (Cohen-Shalev & Rapoport, 1990).

There are simpler explanations, too. A long life filled with experiences can be a source of the old-age style. So, too, are aging's new demands to change; artists cannot work as they once did when they were young. Thus, Wordsworth resisted changing, according to Cohen (1986), because he was unable to shift from the spontaneous and impulsive qualities of his youthful work, where feelings and sensations were emphasized, to the more subtle philosophical and contemplative modes of writing befitting the later years. Consequently, his later poems suffered. Shakespeare, on the other hand, met aging's new requirements by shifting from plays about history to those of comedy and tragedy.

The later works of Tolstoy, H. James, and Yeats also reveal contrasting reactions to aging, according to Edel (1977–78). Tolstoy responded with anger and James with a sense of unfulfillment, but Yeats overcame both and became better. An author achieves a fruitful old age, Edel concluded, when he "finds an expanding power of mind and utterance that can bring him to the supremacies of art... when the artist has endured and suffered and transcended his sufferings, he becomes one of the transcendent beings of art" (p. 68).

Growing old, looking within oneself, becoming more mature, and experiencing personal crises are some ways to account for the old-age style. But they are insufficient. How come the late-life creativity of all aging artists does not molt into an old-age style? A critical trigger is needed, perhaps a particularly severe crisis or conflict, combined with other events. It could be as simple as becoming a grandparent or a widow, or experiencing a long string of successes or failures. A waning of materialistic ambitions or a decreasing concern with prestige could also be a key determinant. Some aging artists may decide they want to express themselves truly, honestly, not as others expect or for the sake of some imagined posterity.

A sophisticated view of the kinds of reevaluations that could take place late in life and set off an old-age style is Erikson's (1963, 1982) model of the stages of life. Elderly artists experience an "identity crisis" or a "generative phase." They look back on their work, evaluate it, consider future prospects, and then make one of several crucial decisions: Stop working, keep working in the same way, or in the case of the old-age style, take a new direction. Old age is therefore a time when artists integrate conflicts, say between hope and despair, and these are reflected in and resolved by sharp and unexpected changes in their paintings, writings, and music (Schaie & Willis, 1986; see especially p. 466). If the outcome is radical, as in the old-age style, it is "a surprise" to everyone, the artist as well as others (Pruyser, 1987, p. 425).

Perhaps the simplest reason for change was put forth succinctly by Goethe. "When one is old one must do more than when one is young" (quoted in Nelson, 1928, pp. 110–111; see also Benn, 1955, p. 299). Put another way, artists react to their own art and feel a need to break away from earlier styles. Hood (1977–78) discussed the need to be different among Renaissance artists who "perceive[d] their late years as something special [because it was] a challenge to continue...indeed, to surpass...their earlier achievements" (pp. 4 and 39). Consequently, the older artists of the Renaissance created an old-age style in "pure" works of art for their own pleasure and benefit. In a similar vein, Simonton (1986) argued that aging artists need to re-prove themselves to a younger generation of artists as well as the public. Evidence for a motive to be different was found among nearly 2,000 classical last works by 172 composers who, at the end of their lives, attempted to make a final and lasting statement, a "swan song" (Simonton, 1989, p. 42).

> Composers in their final years seem to concentrate on producing masterworks that will permanently establish their reputation, doing so by creating works of a concise directness, as revealed by the brevity and melodic

> simplicity of their concluding pieces. [There is] more of an expression of resignation, even contentment, than despair or tragedy.

Echoing Erikson, Simonton concluded that artists retrospectively and prospectively assessed their work as they looked backward over their lives or forward (to their deaths), and asked, "Where have I been?" "What have I been doing?" and "Where am I and my work going?" As creative individuals become increasingly aware of death, Simonton speculated that they

> might adopt a different attitude toward their efforts when they come to realize that among works-in-progress may lurk a last artistic or intellectual testament. As people approach their last years, they may undergo a life assessment, a reflection of how little left remains ... and so may feel that the limited future must be explored to the utmost. For creative individuals, the outcome of this life review may be a significant reshaping of the content and form of those works selected as the career's coda, rendering them qualitatively distinct from other works.

An increasing awareness of death and its impact on the old-age style was echoed by Kastenbaum (1991b). Aging artists responded to "a sense of time urgency" (p. 8). The approach of death can have mixed effects, countered Jaques (1965), in his discussion of Dante's *Divine Comedy*. A sense of one's mortality could explain the emergence of an old-age style but also a decline in creativity. Thus, Winkler (1992) interpreted the increasing consciousness of death in Western art as resulting in works that were either grotesque or wise. Statements about death, many by notable literary figures, suggest other sequela of growing old that might affect the appearance of an old-age style (Booth, 1992).

Many explanations of the old-age style, if not most, along with its definition and description, revolve about the role of illness, physical, sensory and motoric handicaps, and mental debilities like depression, alcoholism, and suicidal tendencies. This doleful list should sound familiar since it also explains declines in creativity, losses in late life generally, and the broad consequences of aging (Chapter 3). There is no reason to believe these negative factors will fare any better in explaining the old-age style than they were for other aspects of aging. It is also inconsistent to argue that poor physical and mental health lead to losses in creativity early in life but result in gains in creativity when applied to the old-age style. It is also unsettling to base an explanation of a high point in creativity, the old-age style, on pathology and debility.

Social, cultural, political, and economic influences are other possible additions to the many explanations of the old-age style (Simonton, 1984b, 1984c, 1986, 1988a). Revolutions, wars, depressions, disasters, and other major events between youth and old age might spark marked changes in the ways artists think and work, and these translate into an old-age style. On a smaller stage, other likely explanations include changes in family circumstances, the influence of colleagues, and altered work habits.

Other explanations might be found in interviews of living artists and biographies of historical artists. But these reports and records are not always available, truthful, or reliable; and different accounts disagree with one another. Diaries, letters, and other first-hand accounts also vary in their inclusion of and emphasize on notable events in an artist's old age. Even so simple a task as establishing the birth and death of artists from the past, in order to date their early and late works, is fraught with difficulties. Artists born before the 16th century or so did not sign their works. Even when signed and dated, there is another major ambiguity: Artists may work on a piece for years, beginning in their youth, but consider it complete only when they are old, at which time they sign and date it.

Summing Up. Explanations of the old-age style encompass a bewildering array of the extraordinary and ordinary, the physical and the mental, the personal and the experiential, the interpersonal and the cultural, and the positive and the negative. There are therefore more than a sufficient number of explanations of the old-age style. But which one—or more likely, which ones—are better? Mumford's (1984) reaction to the similarly abundant accounts of the course of creativity (Chapter 3), many of which duplicate explanations of the old-age style, apply equally well here and bear repeating. "No hypothesis has been proposed which can provide an adequate explanation" (p. 229). Additional explanations are not needed, but rather, fewer and better ones. How do the experts react to this unsettled state?

REACTIONS BY SCHOLARS AND SCIENTISTS
TO THE STATUS OF THE OLD-AGE STYLE

Some art historians are troubled by the many unresolved questions and multiple answers that characterize the topic of the old-age style. The absence of generalizations concerned Rosenthal (in *Man and His Years*, 1954, pp. 12–13). "Do the late works by great artists," she asked rhetorically, "have any particular stylistic characteristics in common (e.g., are they more brilliant or more subdued, simplified or more complex, compared to their earlier works)? Do emotional or dynamic qualities," she wondered, "increase or diminish in later works? Does experimentation continue [or is there] an extension of earlier [tendencies]?" Rosenthal's answer was a faint "not sure." "We have been unable to find any generally valid answers." There are not many studies, she continued, that offer "a general evaluation of artists' early styles." She was not optimistic about a solution, either. There are "few problems in art history that pose more difficulties and negate any general solutions."

Similar complaints were raised by Kemp (1987) when he declared, there is "no simple or consistent thing as an old-age style" (p. 108). Held (1987) also wondered if, "There is not and cannot be, one set of criteria that will do for all artists and all periods to define the style of their old age" (p. 129). He had reservations, too, about the value of studying the single case. It is necessary to look at a large sample, he insisted, in order to determine whether changes attributed to aging in artists are characteristic of all human beings. Rosand's (1987) lament was similar. Although the old-age style is "a commonplace [idea] in art history thought, [it] is hardly a concept that has been subjected to sustained serious examination, nor is it a phenomenon that has been adequately defined" (p. 91). Art historians and literary critics also bemoan, in similarly disappointed terms, the absence of a general description (Brinkmann, 1925; Edel, 1977–78).

The majority of art historians, though, do not seem to be disturbed by the absence of definitive answers to a host of unresolved questions. Inconsistencies, contradictions, and disagreements in discussions of the old-age style represent "richly textured" views. Disparate explanations of the old-age style, the abundance of unanswered questions, and the many unintegrated voices are seen as healthy signs of diversity, the vigor of the discipline, and the uncommon intricacies of the phenomenon. Idiosyncratic positions are valued for their staunch individualism, keen insights, and compelling narratives, not for their consensual, accumulative, and systematic virtues. The apparent indifference of scholars reflects the humanities' distinctive emphasis on the uniqueness of each aging artist and work (Chapter 2), but in this case, it is the scholar whose individuality is championed. In order to maintain the uniqueness and humanness of the old-age style, art historians, literary critics, musicologists, and others in the arts express their individuality by focusing on one or two artists and a few particular works. They do this with case studies of singular examples in depth, and in an unabashedly subjective way.

This kind of particularity is illustrated in a special issue on the old-age style (Rosand, 1987), where each contributor gave extensive attention to one artist or work: Michelangelo and Titian's "Pieta's,"

DeKooning's "Untitled XIV" (1981), two of Rembrandt's "Christ's," and some of the older work of Leonardo, Bernini, and Picasso. Left unanswered, though, were the broad questions raised in the introduction to the issue by the editor (Rosand, 1987) and the summation at the end of the issue (Held, 1987): What are the determinants of the old-age style? Does it exist in other forms, in architecture, as well as in paintings and sculpture? Is the old-age style a "basic aging process of all human beings?" (p. 127). What is the relation between the artist's life and style? And most critically, is there an old-age style? To this reader, at least, none of these questions were addressed by the contributors. But even if they were, general issues raised by the editor and commentator above cannot be addressed by an analysis of one or two artists or a few works with an old-age style. It can be done, though. A notable exception is Cohen-Shalev's (1993), in-depth analysis of a single work by Mondrian which he related to larger concerns about the old-age style. Generality is typically achieved, though, by examining a large number of artists and many examples of their work, which is the scientist's but not the scholar's way.

Scholars are skeptical and wary about (and weary over) the presumed virtues of generalizations based on tedious counts, masses of data, and arcane statistics. Art historians are not upset by inconsistencies between individual artists. Munsterberg (1994), for example, contrasted the old-age styles of the aging Bonnard, DeKooning, Matisse, Michelangelo, Picasso, Rembrandt, and Turner without indicating what they share.

Art historians' reluctance to generalize probably stems from their appreciation of the fact that not all aging artists achieved an old-age style, and that some aged artists, like Grandma Moses, who began painting late in life, are difficult to categorize. Further, the late-life changes of some artists (Chagall, Chirico, Grosz, Munch, Picasso, Utrillo, and Weber) reflected a loss of creativity as they grow older. Then there are the artists whose creativity declined before they became old (Cassatt, Duchamp, and Kollwitz). The late works of other artists, like Monet, fell out of favor in their own time but were rescued by posterity. Hence, an unwillingness to make broad statements about the old-age style. Yet, Munsterberg (1994) made a sweeping conclusion about the sad fate of most older artists. "[E]ven if they have enjoyed a degree of success earlier in life, [they] are neglected or forgotten, displaced by younger people with new ideals and new modes of expression," to which he adds changing fashions of taste (pp. 64–65).

For the scientifically oriented, though, a consideration of a single artist or a handful, or one or a few of their works, are insufficient.

Discursive arguments, informed opinions, and expert judgments based on singular examples of the old-age style may be clever, compelling, admirable, and brilliant. But from an empirical perspective, narratives about individual artists and specific works do not add up to a broad understanding of this phenomenon. Interesting personal reflections are not proofs and cogent explications are not necessarily true. No matter how stimulating scholarly assertions about the old-age style of an artist may be, and many are, if they are of uncertain reliability, questionable validity, dubious generality, limited relevance, and resistant to independent verification by others, they are of limited value. A harsher judgment is Cohen-Shalev's (1989b). After reading more than 50 years of discursive writings on the old-age style, he concluded, "There has, indeed, been little study to back up animated discussions based on prejudice rather than fact" (p. 25). To his credit, he is quick to point out similar shortcomings in scientific approaches to old age which fail to recognize the pertinence of the old-age style to aging, cognition, and creativity. "There has yet been no attempt to relate the old-age style to the life span" (p. 34).

Scientists and scholars' mutual indifference to each other's contributions to the old-age style stems in large part from interdisciplinary barriers rooted in language and training (Chapter 2). In the case of the old-age style, additional difficulties arise from disagreements among scholars over a range of basic issues discussed in this chapter, beginning with inconsistent definitions and divergent descriptions. These uncertainties make it hard to know where to start. With which artists, what works, and how many descriptors should research begin?

TOWARD A SOLUTION: SELECTING ARTISTS AND DESCRIBING WORKS

A good starting point for investigating the issues associated with the old-age style is to reach some consensus on artists with an old-age style. When some authorities choose one set of artists and others select different ones, clouds of descriptions, definitions, and explanations arise, depending on the choices made. With neither a commonly agreed upon set of artists with an old-age style (and those without) nor a workable consensus on good examples, and without a modest accord on their description, advocates of different views of the old-age style can pick and select whatever suits their position. The outcome? Endless and fruitless debate that fails to answer such basic questions as: How many artists had an old-age style? Did artists with an old-age style suffer from physical and other handicaps?

Two fundamental issues are investigated in the following chapters of Part IV: (1) the identification of artists with an old-age style and (2) the selection of a standard set of terms that describes this type of art. Identification and description are at the heart of the many issues that plague this topic. With some certainty about "who" the artists are with an old-age style and "what" describes their work, questions about "why" can be better addressed, along with other matters, such as the difference, if any, between contemporary and historical artists, and between artists and nonartists. But the first step in uncovering the many perplexities of the old-age style begins with the identification of artists with an old-age style and a description of works done in this manner.

The investigations in the next two chapters combine the methodological rigor of scientific inquiry, the expertise of art historians, and the personal testimony of aging artists in experimental studies, surveys, and questionnaires. The success of this endeavor, and its promise in advancing other issues, including interdisciplinary study, are evaluated in Chapter 14.

Identifying Artists with an Old-Age Style: Contributions from Experts, Laypersons, and Artists

> I believe that when one is young, it is the object, the outside world, that fills one with enthusiasm—one is carried away. Later, it comes from inside: the need to express his feelings urges the painter to choose some particular starting-point, some particular form.
>
> The painter Bonnard at age 66, quoted in
> Simone de Beauvoir, *Old Age*, 1972

Progress in understanding the old-age style depends on identifying artists who showed this change (Chapter 11). For the selections to be reliable, the criteria for choosing certain artists have to be spelled out, including those *without* an old-age style. With these broad guidelines in mind, artists with an old-age style were identified in two studies, and their distinguishing characteristics investigated in a third. (Details are in Lindauer, 1993a; Lindauer, Orwoll, & Kelley, 1997.)

ART HISTORIANS' SELECTION OF ARTISTS WITH AN OLD-AGE STYLE

Twelve professors of art selected artists with an old-age style from lists of up to 225 well-known Western painters who lived to at least age 60. (Because this study is unpublished, it is reported in some detail, a practice followed for unpublished research in subsequent chapters.) Artists chosen by a substantial number of experts identify which ones had an old-age style, while those selected by only a few historians or none did not have an old-age style. This consensual criterion is widely used to resolve thorny issues without clear-cut boundaries or certain answers. For example, the frequency with which individuals and works were mentioned in authoritative sources defined their creative status (Lehman, 1953); masterpieces of art were defined according to the number of times they were cited in several collections (Chapter 7); and editors of journals, critics, and textbook writers chose creative architects, mathematicians, and writers (MacKinnon, 1962).

The art historians selected from one of two lists of artists. The shorter one contained the names of 148 long-lived artists who were discussed in an art history textbook (Hartt, 1985). They are the "giants" of the art world; their names would be familiar to most. Upon the advice of several art historians, the list was expanded to include more American artists and a few painters who were also sculptors (e.g., Bernini). The expanded list of 225 artists is therefore a fairly complete set of most of the Western world's recognized artists. Artists from very early time periods (i.e., the 14th century and earlier) were excluded, though, because their birth and/or death dates were unknown or uncertain. In a few cases where the estimated ranges were not too disparate (less than about 5 years), the midpoint of the span was used as the best approximation. The lists therefore included renowned artists from the 15th century to the present; 17 were still alive at the time of the study. A few artists who died before the age of 60, like Pollock, were also included since some art historians claim they had a late-life style comparable to the old-age style of long-lived artists. Dates and spelling of names were checked against the 25 volume *Encyclopedia of World Art* (1968); for contemporary artists, several sources were used (e.g., *Who's Who in American Art*, 1990).

The shorter list of 148 artists was distributed to 18 art historians attending an institute on contemporary theories of art; eight useable booklets were returned (44%). The expanded list of 225 artists was mailed to eight art historians on the faculty of a large midwestern university; four useable surveys were returned (50%). The 12 historians were instructed to check the names of artists whose "style significantly changed in old age" and to place a question mark next to artists about whom they were uncertain. Nine of the 12 historians who returned the lists (46%) also completed a background questionnaire: There were slightly more women than men; their specialties were historical and modern European and American art; and they all had PhDs in art history, were full professors, and had extensive teaching experience (Mean years = 15.78; Median = 15.00).

Artists chosen by at least a third of the judges (four or more) were defined as having an old-age style. The criterion was set relatively low because the historians' choices were extremely variable. For example, 5, 11, 76, and 122 artists were selected by four judges; and no artist was unanimously chosen by all 12 judges or even by 11. The lack of unanimity reflects the difficulties of judging the old-age style (Chapter 11). Had a stricter criterion been used (and undecided votes disregarded), the opinions of many judges would not be represented.

In terms of the criterion, 38 artists (17%) had an old-age style: Titian (selected by 83% of the judges, the highest number); Cézanne, Constable, Degas, DeKooning, and Goya (75%); Matisse, Michelangelo, Mondrian, Monet, Picasso, and Rembrandt (67%); and Braque, Corot, David, Kandinsky, Kokoschka, Renoir, and Turner (50%). Receiving less than a majority of the votes, but still meeting the criterion, were Corinth, El Greco, Hals, Klee, Nolde, Poussin, and Rothko (chosen by 42% of the experts); and Bacon, Botticelli, Chagall, Chirico, Dali, Delacroix, Derain, Eakins, Fragonard, Gainsborough, Homer, and Vuillard (33%).

Sixty-eight artists were not chosen by a single judge, and therefore clearly did not have an old-age style. They included such well-known artists as Albers, Bingham, Cole, Giotto, Gorky, Orozco, Peale, Prendergast, Man Ray, Soyer, J. Strella, Trumbull, Uccello, Van Dyck, Warhol, and A. Wyeth. Another 92 artists received only one vote and were also considered bereft of an old-age style. Among the familiar names were Bernini, Cassatt, Courbert, Davis, Frankenthaler, Gris, Innes, Lichtenstein, Magrittte, Manet, Millet, Miro, Neel, Nevelson, Piazetta, Pollock, Rauschenberg, Reynolds, Ruisdael, Shahn, Stuart, Toulouse-Lautrec, Velasquez, Vigee-Labrun, Weber, and West. An additional 39 artists, selected by only two judges, were unlikely candidates for an old-age style. A partial list included Bellini, Canaletto, Chardin, Dubuffet, Duchamp, Durer, Ernst, Gauguin, van Gogh, Ingress, Johns, Kollwitz, Marin, Ryder, Tiepolo, and Utrillo. Twenty-nine artists, selected by three judges, fell just one vote short of criterion. These 29 marginal artists warrant a second look as possible candidates for an old-age style. Some high-profile names were Benton, Bonnard, Copley, Friedrich, Hopper, Kirchner, Leger, Munch, Murillo, O'Keefe, Pissaro, Rouault, Sargent, and Tobey.

The 38 artists chosen by the panel of art historians were compared to three major sources on late-life art: *Man and His Years* (1954), Munsterburg (1983), and Rosenthal (1968). Not unexpectedly, little agreement was found across the four (the panel of experts and the three published works). Only 15 of the 225 artists in the survey were mentioned in the three published sources, and of these 15, the status of seven (47%) was the same in each. An old-age style was unanimously assigned to Cézanne, Goya, Michelangelo, and Titian. These four might therefore be defined as prototypical examples of the old-age style because of the unanimity of the choices. On the other side of the coin, none of the four sets of experts attributed an old-age style to three artists: Chardin, Hartley, and Manet. The results were mixed for eight artists (53%). Corot, Degas, El Greco, Homer, Marin, Pissaro, Renoir, and Rubens either did or did not have an old-age style, depending on the authority consulted.

The choices by the panel of experts were compared to one of the three collections (Rosenthal, 1968) because it included more artists who were also in the survey (39) than the other two. The panel and the collection agreed on 14 (36%): Braque, Cézanne, Fragonard, El Greco, Gainsborough, Goya, Homer, Kandinsky, Monet, Rembrandt, Renoir, Turner, and Vuillard had an old age style and Reynolds did not. The two sources disagreed, though, on the remaining 25 artists (64%): Beckman, Bonnard, Cole, Corot, Gauguin, Guardi, Hals, Hopper, Innes, Kirchner, Manet, Miro, Peale, Pollock, Rubens, Ruisdael, Sargent, Strozzi, Stuart, Tobey, Toulouse-Lautrec, Van Dyck, Villon, West, and Whistler.

There was therefore more disagreement than agreement between the panel of historians surveyed and the three leading published authorities, which was not surprising since the latter three also disagreed with each other. What was certain were the 38 historical artists selected by art professors as having an old-age style and at least 68 who did not. The advantages of establishing a "certified" or standard list of artists with an old-age style for future research are discussed subsequently.

LAYPERSONS' JUDGMENTS OF YOUNG–OLD PAIRS THAT IDENTIFY ARTISTS WITH AN OLD-AGE STYLE

Sharp distinctions between an artist's youthful and late art is another way of identifying those with an old-age style. Young–old pairs that differ the most are likely to have been painted by artists with an old-age style; by their works shall they be known. Thus, a work done in the old-age style would seem to be by an artist other than the author of an earlier work, while late works by artists without an old-age style should be indistinguishable from their younger efforts in brushstroke, detail, and the like. The problem is determining what differences to look for, since art historians discuss so many and disagree (Chapter 11). A rougher brushstroke in older works? A more hurried appearance? An unfinished look? Subject matter? There are too many to try all of them but which ones to begin with?

Four possible distinctions were chosen. Earlier and later works by artists with an old-age style, compared to artists without one, should (1) look dissimilar in style, so much so that (2) they seem to have been painted by an older and (3) a different artist, and (4) have less detail. Differences in style, age of production, authorship, and detail carry a certain obviousness, commonsense appeal, and historical credibility. Distinctions between young and old paintings on these four dimensions should therefore identify artists with an old-age style, while pairs that differ on a few or none of these would point to artists without an old-age style.

Another set of critical methodological questions: Which early and late works of art, by which artists, should be compared, and how many examples? On these matters, as on most about the old-age style, authorities are of little help (Chapter 11). The problem was met by relying on a single authoritative source for the appropriate artists and best examples of their youthful and aged works. Luckily, the Rosenthal (1954) collection on late-life art illustrates 72 pairs of early and late works by 70 long-lived artists who were selected and discussed by 30 experts. (Two artists, Matisse and Picasso, were represented by two pairs.) The experts, as elsewhere, were not free from the uncertainties that plague this topic. For example, there was some ambiguity about the old-age style of seven artists: Bellow, Dubuffet, Hals, Klee, Picasso, Reynolds, and Tobey.

Nevertheless, despite some drawbacks, the Rosenthal collection has several advantages. The choice of artists, and of prototypical examples of their younger and older work, were based on the judgments of specialists in art history drawn together for a common purpose, in this case, an exhibition. In addition, the experts worked under the direction of a single editor/curator. It seems reasonable to assume that difficult and controversial decisions about which artists and works to include were presumably discussed under some set of guidelines, albeit these were not made explicit, and with definitional and other issues minimized if not completely resolved by a consensus of a large group of informed scholars (again, a point not mentioned).

From the larger set of 72 young–old pairs, 24 by different artists were selected on the following basis: Both members of the pair had a similar subject. That is, both the young and old members of a pair depicted a landscape, portrait, and so on. Had subject matter radically differed between a pair, then differences in content as well as style would play a role in the judgments. The matching, unfortunately, meant that some artists with a reputation for an old-age style could not be included. Rembrandt, for example, was a major loss. (His youthful work was a group portrait, *The Tribute Money* (1629), while his older work was a portrait of an individual, *Titus, the Artist's Son* (1660).) Excluded, too, unfortunately, were other artists who are known for their old-age style (Cézanne, Goya, Michelangelo, and Titian).

The younger members of the pairs were painted when the artists, in general, were in their 20s to 30s, while the older works were done when they were past 60 (the average age of death was 71.54). Five relatively short-lived artists were included because the experts in the Rosenthal collection discussed their old-age style. Of these five, Cole, Klein, and Manet were in their late 40s to early 50s when they died, and two, Kirchner and Klee, were in their late 50s.

The next critical procedural question: Who should judge the 24 pairs? Untrained laypersons were chosen. They represent the majority of attendees at art museums, and as unsophisticated viewers they have several advantages over experts. Their judgments are based on what they *see* rather than on what they *know*. Untrained judges were therefore more likely to be more spontaneous and open than experts, and less concerned about being correct. In contrast, judges knowledgeable about art history (professors, majors, artists) were likely to evaluate a work of art on the basis of what they *know* about the painting, the artist, and the old-age style, and the views of other authorities. A sophisticated viewer, when shown a Rembrandt, for example, might reason, "I recognize this work as a late Rembrandt, he is known for his old-age style, and his work became less detailed as he grew older. Therefore, I choose this work to exemplify an old-age style." Neutral observers, in contrast, make unbiased judgments based on perceived differences between early and late works.

Consequently, four different sets of untrained judges rated or ranked the pairs on four tasks. Pairs judged to be the most or least different in style, authorship, age of production, and detail were probably painted by artists with and without an old-age style, respectively. An example: If the older member of a pair by Monet painted in the artist's late 60s were judged different in style from a work done in his 30s, and also appeared to be done by an old artist (age of production), and indeed, was not believed to have been painted by Monet as a young man (authorship), and in addition, seemed to have less detail than its earlier counterpart, then this work has an old-age style and Monet is so identified as an artist who changed. In contrast, if the pair by a youthful and aged artist Tiepolo in his 20s and 60s looked similar in style, as if both examples were done by the same artist (as in fact they were), with neither member of the pair looking as if it had been painted by either an older or younger artist, and moreover, the two appear to have the same amount of detail, then Tiepolo's pair (and he) did not have an old-age style.

In addition to the 24 young–old pairs, eight pairs by different artists were also shown; one member of the pair was painted in an artist's youth and the other by a different and old artist. Thus, the works in these eight young-old pairs were distinct from one another by definition. An example: A young work by the American artist Whistler (1834–1903), *At the Piano*, painted around 1858 when he was 25, was paired with an old work by the French artist Corot (1796–1875), *The Crown of Flowers*, done between 1865 and 1870, when he was about 69. Judgments of the authorship and other three dimensions of the eight different-artist pairs should be similar to the judgments of the same-artist pairs with an old-age style, and both should be dissimilar from pairs not judged to have an old-age style. Thus, in the example above, if the young–old members of the Monet pair were distinguished by an old-age style, they should look as different from one another as the pair composed of a youthful Whistler and an aged Corot.

The pairs were a fair sample of the world's art, representing a wide variety of artists, types of art, nationalities, and time periods. The 24 same-artist pairs included 11 American artists, four Frenchmen, two each from Italy, Germany, and Spain, and one each of British, Swiss/German, and Dutch nationalities. The 20th century was represented by eight artists, six bracketed the 19–20th centuries, four were painted during the 19th century, three fell between the 18th and 19th centuries, and three were done in the 18th century or earlier. Three kinds of art composed the pairs. Two were representational (realistic): figurative works by Corinth, Eakins, Goya, Manet, Reynolds, Sargent, Stuart, and Tiepolo where people predominated (portraits, groups, genre, religious, mythological, and historical works); non-figurative paintings by Bellows, Copley, Cole, Guardi, Innes, Kirchner, Monet, and Pissaro composed of land- and seascapes and still-lifes in which people were either absent or minimal; and non-representational (abstract or modern) art by Hoffman, Klee, Klein, Leger, Marin, Mondrian, Picasso, and Tobey. The three kinds of art were equally represented in the 24 same-artist pairs, eight of each kind, and as evenly distributed as possible among the eight different-artist pairs by Beckman/Bonnard, Cèzanne/Braque, Delacroix/Hopper, Matisse/Gorky, Miro/Gris, Turner/Piazetta, West/Cassatt, and Whistler/Corot (the younger work is listed first). The 24 same-artist pairs were as similar as possible to the eight different-artist pairs, in time periods and nationalities, as were the intra-pair characteristics of the latter, although this was not always possible, given the limited sample of 70 pairs to work with. None of this information, especially the identity of the painters and the nature of the pairs, was available to the judges. An additional eight pairs were used for purposes of training and warm-up. Thus, the judges saw a total of 36 pairs. Slides of the pairs, shown simultaneously, like the reproductions in the Rosenthal catalog from which they were photographed, were in black and white. (For a discussion of the differences between originals, slides, prints, and black and white photographs in doing research with art, see Dreher, 1968.)

The pairs were subject to four (and in some cases, three) sets of judgments. On the identification task (authorship), each pair was rated with a 5-point scale on whether it looked as if it had been done by the same artist or not. Ratings were also used to indicate the similarity in styles between the members of the pair. For age-of-production, judges chose the member of the pair that looked as if it had been done by an older artist. On judgments of detail, viewers chose the member of the pair with less detail. The detail task differed from the other three in several ways: The slides were presented at the beginning of a lecture on old-age art to a small group of older participants (nine compared to 22–24 in the other tasks). The group was nearly equally composed of graduate students in psychology and professors from several disciplines. Consequently, the participants most likely had more experience with art than the undergraduate judges. Further, the number of pairs they saw was reduced to 15 (rather than 36) because of time constraints.

A pair met the criterion for having an old-age style when judgments about its authorship and the like statistically differed from the other pairs on at least three of the four tasks, or in the case of the detail task, where some pairs were omitted, two of the three tasks. Pairs differing in only one way or none across the three to four tasks were considered to be by artists without an old-age style. The procedures and criteria are more concretely illustrated in the results that follow.

Fifteen of the 24 pairs (63%) were sufficiently distinguished to be called works with an old-age style (Lindauer, 1993a, table 1, pp. 140–141). One pair differed on all four tasks (meeting 100% of the criteria): the American Eakins' (1844–1916) *Home Scene* (1870–72), painted when he was about 28, and *Portrait of Mrs. Edith Mahan* (1904), completed when he was about 61. Distinguished on three of four tasks (75%) was the pair by the French artist Monet (1840–1926), represented by *Garden of the Princess, Louvre* (c. 1867), painted at about age 35, and *Water Lilies* (1914), produced when he was about 75. Noteworthy, too, were four pairs by Kirchner, Mondrian, Guardi, and Klee, which were distinguished on two of the three tasks (66%) on which they were judged (they were not included in the shortened detail study). Meeting the criterion, too, on half of the four judgmental scales, were the nine pairs by Bellows, Cole, Goya, Innes, Picasso, Pissaro, Reynolds, Sargent, and Tobey. Most of the works with an old-age style were landscapes (seven); the remainder were equally divided between figurative and non-figurative examples (four each).

Nine pairs were not distinguished on three or four tasks (the number varied, depending on whether the pair was included in the detail task). Five pairs by Corinth, Copley, Leger, Manet, and Stuart differed on only one of three tasks (33%), followed by Klein's pair, which was distinguished on only one of four tasks (25%). Three artists failed to reach criterion on any task: the American non-representational artist Marin (1870–1953), whose works were *East River* (1910), painted at age 41, and *The Tempest* (1952), done at 83; the Italian Tiepolo (1696–1770), with the figurative pairs, *Susanna and the Elders* (c. 1718–20), produced when he was about 24, and *Venus and Valean* (c. 1755–60), painted when he was 63; and the American Hoffman (1880–1966), whose abstract work *Fantasia* (1943), done at age 64 (yes, the Rosenthal collection identified this work as an example of his "younger" work) and *Summer Night's Bliss* (1961), painted at 82. Works without an old-age style, with one exception, were equally represented by figurative and abstract art (four each); the exception was the pair by Stuart, a non-figurative landscape.

Detailed results for two artists are illustrative. Among non-figurative representational pairs, the older work by Monet met three of the four criteria: 62% of the judges judged his older work to be painted by an older artist; and mean ratings of the pair were statistically higher than the others on authorship, indicating that the judges thought the two were done by a different artist (Mean=3.80), and were judged to be different styles (Mean=3.91). However, none of the judges judged the pair as sufficiently differing in detail. The four scores for Monet were combined into a z-score, an index that transforms percentages and ratings into a common value. On this basis, the pair by Monet was more distinctive than others (z-score = 4.48). In contrast, the figurative pair by Tiepolo met none of the criteria: his older work was judged as different by only 17% of the judges, and authorship and style ratings were statistically undistinguished (Mean = 2.56 and 1.41, respectively). (The pair was not included in the detail task.) This pair's singular lack of distinction was reflected by the lowest z-score of any artist (= −5.21).

The 15 pairs with an old-age style were distinguished in another way. As a group, their judgments were more similar to the eight different-artists pairs than to the nine same-artist pairs without an old-age style. For example, on similarity of style, the nine same-artist pairs without an old-age style were statistically different from both the same-artist pairs with an old-age style and the different-artist pairs; the latter two did not (statistically) differ from one another (Mean ratings = 2.40, 2.95, and 3.40, respectively; the lower the score on this scale, the higher the similarity). The identification of authorship followed the same pattern: The pairs with an old-age style and the different-artist pairs were more often judged as painted by different artists than the pairs undistinguished by an old-age style (Mean=3.43, 3.40, and 3.05, respectively). A similar trend characterized the identification of age: The older members of the 15 pairs with an old-age style and the eight different-artist pairs were more often correctly chosen as older by almost the same percentage of judges (Mean percentage chosen=42.20% and 42.88%, respectively), and both differed from the nine undistinguished same-artist pairs (Mean=31.44%). The combined z-score also reflected the patterns found on each of the four tasks considered separately. The undistinguished pairs by nine artists (Mean z-score=−1.41) statistically differed from the 15 artists whose works indicated an old-age style (Mean z-score=0.84) and from the eight different-artist pairs (Mean z-score=0.00).

The Status of Uncertain Artists

The contributors to the Rosenthal collection, from which the pairs for the study were drawn, disagreed with the editor on the amount of change in the later works of Bellow, Dubuffet, Hals, Klee, Picasso, Reynolds, and Tobey. The judges in this study were also uncertain about these artists, as indicated by the mixed judgments given to these artists' works. Thus, the identity of these seven artists' pairs (the authorship task) was about as accurate as the other 17 artists, but judgments about their style differences were inconsistent, compared to the rest of the pairs, and proportionately more of the seven questionable pairs looked as if they were done by old artists, compared to the other pairs. Mixed judgments, too, characterized the five artists who died at a relatively young age but were considered by the experts in the Rosenthal collection to have an old-age style. Pairs by three artists (Cole, Kirchner, Klein) were judged to have an old-age style across four tasks while two (Klein and Manet) were not.

The Number of Artists with an Old-Age Style

Two sets of data on the old-age style, the judgments of pairs in the Rosenthal collection of late-life art, and the selections by the panel of

art historians from a worldwide list of aging Western artists, are a basis
for estimating the number of historical artists with an old-age style.
The calculations proceeded as follows: The undergraduate judges
selected 15 of 24 pairs as having an old-age style, or 63%. When this
percentage is applied to the entire collection of the late works by 70
long-lived artists, as many as 44 artists could have an old-age style.
(Recall that at least five artists known for their old-age style—Cézanne,
Goya, Michelangelo, Rembrandt, and Titian—were not included in the
study because of subject differences between their pairs, and for some
contributors to the collection, at least another 12 [Bellow, Dubuffet,
Cole, Kirchener, and others] could have an old-age style.) This number
(44), as a percentage of the 225 recognized Western artists, is about
20%, a figure not too different from the experts, who selected 38 artists
from 225 well-known Western artists, or 17%.

There is therefore an empirical rationale for estimating that about
one fifth of Western artists probably had an old-age style. The number
is certainly high enough to support the argument that an old-age style
exists among historical artists. Further, the number is not so large as to
make detailed case studies of artists with an old-age style impossible,
yet large enough to provide a suitably large sample for investigations of
this type of late-life creativity.

Future Research

A good place to begin research would be the artists most likely to
have an old-age style. Meeting this ideal were the 12 artists whose
works were judged by untrained judges as having an old-age style *and*
who were also among the 38 artists selected by the panel of art historians. The 12 artists, fortuitously, were equally divided between those
with an old-age (Eakins, Goya, Klee, Mondrian, Monet, and Picasso)
and a "control" group who did not (Copley, Hoffman, Manet, Marin,
Stuart, and Tiepolo).

Armed with this list, scholars can compare the six artists who
changed with the six who did not by examining their autobiographies,
biographies, notebooks, and letters. They might note differences
between the two sets of artists' emotional states, temperaments, and
relationships. The 12 artists with and without an old-age style might
have calm or agitated approaches to life, successful or stormy family
and personal relations, helpful or unhelpful treatments from teachers,
colleagues, and patrons, extreme reactions to fame and failure, youthful
successes or late-life disappointments, positive or negative evaluations

of their later works, and diverse experiences of war and peace. Thus, personal accounts by and about Eakins and Tiepolo, artists markedly with and without an old-age style, would throw light on the kinds of factors that could have affected their work, the freedom to develop an old-age style, the acceptance of their late works, and the need to prove themselves in old age.

Differences between the lives and times of these 12 artists could clarify other contentious issues, such as the impact of physical, motor, and sensory losses on this form of late-life creativity (see Chapter 13). The six artists who developed an old-age style, for example, but not the other six, may have compensated for motoric, sensory, and other handicaps, and this is why their works, and not that of others, were characterized by a rougher brushstroke and an impatience with detail.

A scholarly unraveling of the contrasting profiles for 12 artists with and without an old-age style might also suggest why creativity was drastically altered for six, and continued, remained unchanged, or declined for the other six. Discussions of these matters would guide interviews, questionnaires, checklists, and rating scales of contemporary artists. The outcomes would amplify, support, or revise scholarly speculations.

Contrasting biographical accounts for the two sets of artists might also throw light on whether historical artists with an old-age style, unlike those without one, saw their work and themselves as different from younger colleagues and their paintings, believed their late works were rejected, felt a need to prove themselves in old age, thought of themselves as different from their younger selves, or evaluated their later works as better, different, or the same as earlier ones. Comparisons between artists' youthful and aging selves were the focus of the next study of contemporary artists.

COMPARISONS BETWEEN OLD AND YOUNG ARTISTS

Distinctions between old and young artists might suggest some of the personal, work-related, and environmental differences that set apart those with an old-age style from those without one. To this end, nearly 90 recognized artists in their 60s to 80s who completed a questionnaire on four aspects of their creativity (Chapters 9 and 10; Lindauer, et al., 1997) were asked, "What differences, if any, do you see between yourselves and younger artists?" a comparison that also included the artist's own youth.

Most of the artists (86%) reported that there were indeed differences between young and old artists and works, including themselves as young men and women. A 65-year-old man put it this way: "Of course there are differences, just as there are between the young and the old on anything." However, only about half of the artists (55%) specified what these differences were; the nine most frequent ones follow.

The distinction most often mentioned by over a third of these artists (37%) was increased knowledge, or words to that effect: skills, training, background, experience, thinking, observation, understanding, thoughtfulness. The major distinction between young and old artists and art was once again knowledge, just as it was in accounting for changes in creativity with age (Chapters 9 and 10). Not unexpectedly, the majority (64%) of artists who emphasized knowledge believed older artists had more. "Young artists simply have not had time enough to live, observe nature and think about picture making," said a 71-year-old man. The benefits of increased knowledge with age were spelled out in the following two comments.

> Young people have not had enough primary and direct life experiences, no matter how brilliant their minds are. They have not had time to study, read, absorb, and integrate past accomplishments in history, and by other artists, into their own efforts. (Written by a 66-year-old woman)

> The more one reads, looks at art, learns, listens to great music old and new, relates to others, suffers pain and joy, the more all of this is reflected in the [older person's] work. (Reported by a 68-year-old woman)

A 76-year-old man pointed out, though, that the belief in old age bringing greater knowledge did not necessarily mean better art. "Old artists, by virtue of their experience, know more—but they are not necessarily more talented." In other words, gains in knowledge could be a mixed blessing, a point amplified by a 63-year-old man.

> Older artists have attained some wisdom and depth over the years. But as a young artist, I had the feeling there was no tomorrow and I could achieve anything in my work. There's a good feeling about being a young artist that I miss very much and I can't duplicate as an older man.

Interestingly, approximately one third (36%) of the reporting artists wrote that the *young* had more knowledge. By this they meant that youth had *special* knowledge, the result of being born, raised, educated, and trained at a time that gave them certain advantages over older artists, such as more freedom and increased opportunities.

The second major age difference (reported by 33% of the artists) was a willingness to experiment, that is, to be innovative and open to the new. Old and young artists were reported to be nearly the same on this trait (48% and 44%, respectively). That is, both age groups take chances, and are equally free, daring, spontaneous, uninhibited, experimental, and adventuresome, as well as try new things, do not follow rules, and do not work in a routine way. Some of the favorable comments made by older artists about younger ones follow:

> Young artists want to do something new. They want to establish their identity apart from older artists. Older artists tend to not want to do something new. Only a few break new ground. (A response from a 65-year-old woman)

> Young artists are struggling to find a way to work that they feel good with. Older artists are less experimental, although there are many exceptions, e.g., Picasso and others. (A 76-year-old man)

> Younger artists tend to be more innovative. They were usually taught to be freer. We were pretty hidebound in that good drawing was stressed, and also we were hindered by the "you NEVERS"—and "you ALWAYS." Now you do whatever works for you, and this is hard to learn for the "old-timers." (An 80-year-old woman)

The third frequently mentioned age difference (reported by 21% of the artists) alluded to career orientation. Of those who gave this reason, most (75%) said that younger artists strived more than older artists for reputation and wealth. Older artists were no longer obsessed by dreams of a career, by nagging and driving ambitions, or by pressures to succeed.

> Younger artists, on the whole, are more daring and hurried. They want to attain recognition and wealth while still young. Older artists know the ways of the art world, and know that if they achieve recognition, it will interfere with their more profound feelings and they will not be able to develop to full maturity slowly and surely. (Reported by an 83-year-old man)

> Experimentation at the cost of learning, for the sake of career advancement, is my criticism of younger artists. From what I see and hear, younger artists seem to carry a pathological reticence to emulate anyone who was considered great over 30 years ago. Their goals seem to be innovation and making money! Fools! (A 66-year-old man)

Other distinctions between young and old artists/art were less frequently reported than knowledge, innovativeness, and career orientation, the three major differences discussed above. Energy level was mentioned by 19% of the artists, with most (67%), not unexpectedly, attributing a higher amount to younger artists. Still, a substantial third of the older artists disagreed. Further, some who attributed greater energy to youth did not think it was a virtue. "The younger artists," wrote a 76-year-old woman, "usually express a lot of energy in their work, but as an artist matures he learns to rely upon his knowledge of life more and his experiences in dealing with it."

Surprisingly few artists (16%) indicated that maturity, synonymous with age, distinguished older from younger artists. A 66-year-old man made the case this way:

> "Older is better," is of course a cliché. But I believe it is true in the arts: Greater depth of understanding—of nature, society, and of one's self—comes with age. An age that is alive and alert can't help but make for greater art. The older artist is also less likely to be seduced by the tricks of the market place.

Yet about a third of the artists who specified differences did not consider maturity a necessary outcome of aging or limited to the old. Some also pointed out that maturity did not come automatically without additional qualities, as this 71-year-old woman's comment illustrated.

> Maturity cannot coexist with inflexibility and rigidity, The very young artist is too immature to think things out. It takes years to really assimilate and judge. With older artists, though, the problem is you have to watch out for an inability to try something new or to close down [your] horizons.

Favoring the older artist, as well, was greater self-acceptance (reported by 13% of the artists), a trait often linked with self-confidence. "As a young person I lacked the complete confidence I now have, and it keeps on coming," said an 80-year-old woman. Another 80-year-old woman wrote that self-acceptance meant being able and willing to express oneself. "Older artists produce more genuine expressions of themselves because they have developed skills, and they are freer to choose what they do."

Two other distinctions were noted by a relatively small number of artists (11% each). Creativity as a function of age was the first one. However, the comments were either unclear as to where the age advantage lay or creativity was assigned equally to young and old artists. A clearer age distinction was attributed to the ability to concentrate, at which the elderly excelled. "As you age you get slower, more thoughtful...but you can focus more sharply on fewer things," reported a 77-year-old woman. The last age difference of some note (reported by 8% of the artists) was the belief that old artists mellowed with age, that is, they became less critical of themselves and others.

Summing Up. Seven characteristics distinguished old artists and late art from young artists and youthful efforts. Older artists have more *knowledge* and are *less career oriented*. They also have *less energy*—the only case where older artists were at a disadvantage to younger ones—which they compensated for with *greater maturity*, *concentration*, and *self-acceptance.* Older artists were also *less critical* than their younger counterparts. However, in two areas, *creativity* and *experimentation,* older artists were seen as equal to younger practitioners.

IS THERE A CONTEMPORARY OLD-AGE STYLE?

Most of the notable differences between young and old artists' art were more related to the self (knowledge, career-orientation, energy, maturity, concentration, self-acceptance, criticalness) than to their work (creative and experimental). Changes with age therefore emphasized personal rather than work-related factors, and this may explain why the 60- to 80-year-old artists did not refer to drastic differences in the way they painted as they became older, or more specifically, to an old-age style. Where changes were reported, they occurred gradually rather than took radically new paths. As a 77-year-old woman put it, "As I aged my style evolved, but never really changed." Some mention was made of "old age" but in the context of health rather than to style. A few passingly mentioned the possibility of a special style in old age, only to reject it. For one 66-year-old woman, a drastic change would have been unacceptable for commercial reasons. "If my style changes too much, buyers will reject my work." A 63-year-old man dismissed the idea of stylistic change on artistic grounds: "What's important is developing and striving for some kind of excellence in a particular direction. I've never been impressed with artists whose styles are always changing."

The failure to *report* an old-age style does not necessarily mean the artists did not have one (Chapter 10). They may simply have failed to mention it since it was not specifically asked about. The artists may have been reluctant to refer to old age because of its association with decrepitude, senility, unproductivity, and other stereotypes connected to ageism. The term also hints at the approach of death and the end of a career. Allusions to an old-age style may therefore have been deemed inappropriate for artists who consider themselves vigorous, active, and productive. The artists may have also been hesitant to use the term because of its historical connection to artists whose late-life works disappointed contemporaries, and were criticized for having failed to fulfill their youthful promise. Consequently, for contemporary artists to label themselves as having an old-age style might have implied that their work was unacceptable. On the other hand, the artists might have been embarrassed about touting an old-age style since the term evokes images of such luminaries as Cézanne, Goya, Michelangelo,

Rembrandt, Titian, and other remarkable old painters with astounding achievements. The artists may therefore have been reticent to put themselves in the same category as the greatest figures of art. It is also possible that the artists underestimated the amount and degree of change in their work. Recall that in earlier chapters (Chapters 9 and 10) these artists reported marked increases in four areas of creativity, especially in the quality and quantity of their work, as well as in the sources of their ideas and their approach to art (the latter was a synonym for style). The problem, too, may lie in how they defined "change" and "style," difficulties discussed earlier (Chapter 11).

The term "old-age style" was purposely not used in the questionnaire because of its loaded connotations. Instead, its presence was explored indirectly by asking older artists about differences between themselves and younger artists, a more neutral question. Reports of an old-age style therefore depended on spontaneous comments, and these, regretfully, for the purpose of this study, did not occur.

Self-reports, in other words, do not necessarily give access to innermost thoughts, the absence of comments does not necessarily mean there was nothing to reveal, and what is said is not easy to interpret. Take the artists' comments about young and old artists being equally creative. Yet self-ratings on their creativity consistently showed sharp *increases* with age (Chapter 9, Figures 9.1 and 9.2).

Other kinds of reactions to aging, more obvious than an old-age style, were not reported either. Few if any of the old artists said their work reflected their age or was "typical" of this particular time of life. No one spoke about taking life less (or more) seriously in the few years left to them, of time running out, or of feeling more (or less) urgency or deliberateness as their lives were coming to an end. Similarly, few artists wrote about working harder or faster, of giving their art more attention as death approached, or as the sureness of their hand was lost. The artists did not, at least explicitly, see their last works as final testaments to their lives, as "swan songs." Not many artists were concerned, if we take their written accounts at face value, about becoming more anxious, profound, or challenged as death neared. The idealism of youth or the wisdom of age were hardly mentioned. Remarks about having to work more slowly rarely appeared, although psychologists find a slowing rate of response as a major distinguishing feature of aging (Chapter 15). What old artists did emphasize, in distinguishing themselves from younger colleagues, was increased knowledge—just as they did in accounting for increased creativity with aging.

The old-age style is a rare phenomenon; the best estimate is that it applies to about 20% of the world's most esteemed artists (Chapter 12).

With less than 90 artists responding to the questionnaire, the likelihood of finding the handful with an old-age style was quite small, especially when only the most frequent accounts of young–old distinctions were highlighted (involving 8–37% of the artists). With so few artists having an old-age style, their voices were muted. The nature of the sample, too, may also have reduced the number of artists with an old-age style. Those who declined to participate may be more individualistic and independent than the artists who did volunteer, and it is among the former in whom an old-age style is perhaps likely to be found.

The artists who did volunteer, although judged as creative by their peers, probably included few whose works will be installed in major museums in the future, if the history of art is any guide. But then again, very few contemporary artists and most of the artists who ever lived, have not been enshrined in museums. Even among the handful of today's well-known, not many will remain famous 50 years after their deaths. Cézanne, Rembrandt, and others died in relative obscurity, unacknowledged and unrecognized for their greatness (and an old-age style), but are today held in the highest esteem. The lasting value of an artist's work, as well as the identification of artists with an old-age style, requires some distancing with time. The history of art is replete with artists for whom a considerable number of years passed before it could be said with certainty that they maintained, enhanced, or lost their reputations. In the last analysis, the recognition of discontinuities between an artist's younger and older productions that define an old-age style await the passage of time.

The long-livedness of today's artists also works against the detection of an old-age style. Artists in their 60s will probably live, on average, past their 70s; women longer. Today's improved financial, medical, and living standards, compared to the past when 60 was really old, most likely means that an old-age style will probably emerge much later in life than it did historically, at 75 or so. Thus, the 60-year-old artists, and many of the 70-year-olds who were studied, were "too young" to have an old-age style; and there were too few 80-year-olds to manifest this rare phenomenon.

Even if the old-age style emerges later today than it did in the past, it is unlikely to take a form that is recognizable, compared to the predominantly representational works of the past. Abstract and modern art in general do not depict familiar forms and content. Consequently, many if not most of the characteristic features of the old-age style known from the past, such as the treatment of details and light, are not applicable to today's materials, techniques, and media.

The revolutionary changes in Western art in the last 100 or so years— abstract art, expressionism, colorfield, cubism, dada, hard-edge, impressionism, minimalism, non-objective, op-art, and post-modern— have considerably reduced the body of representational art, where the old-age style is most clearly evident. The old-age style, strikingly different today, would therefore be extremely difficult to recognize among current artists. The question must therefore remain open as to how many of the creative and recognized contemporary artists who responded to the questionnaire had an old-age style, as it is for today's artists.

There are some hints, though. Although contemporary artists did not identify themselves as having an old-age style, the nine major distinguishing young–old differences, in exaggerated form, might suggest what late-life changes might look like. Artists with an old-age style might have the following traits: open to new knowledge, not career oriented, a high energy level, mature, self-confident, focused on their work, and indifferent to their future status. Counterintuitively, though, extrapolations from the questionnaire results suggest that artists with an old-age style might not consider themselves more creative or experimental than artists whose styles did not change.

The nine dimensions that might identify artists with an old-age style can be combined with the 15 major reasons artists gave for changes in their creativity with age (Chapters 9 and 10). These 15 (in decreasing order of importance) were knowledge, self-acceptance, physical change, time to work, favorable circumstances, motivation, understanding of self and others, the priority given to art, technique, maturity, concern for excellence, a willingness to experiment, a shift to or from abstraction, changing family responsibilities, and the exploration of different kinds of subject matter. Delete the four repeated traits from the combined list of nine and 15 (knowledge, maturity, self-acceptance, experimentation) and 20 remain. Their usefulness can be tested in the form of a checklist or rating scale applied to biographical material about the 12 historical artists noted earlier as prototypical examples of the old-age style or its absence. How much concern did the two sets of artists show to criticism, the judgments of others, and the evaluations of critics? Did they do art for its own sake, were they willing to go their own way, and so on? Signs of heightened independence, increased autonomy, greater freedom, and other qualities tapped by the list of 20 distinguishing characteristics might successfully differentiate artists with and without an old-age style. As a further check of their validity, the 20 characteristics could be applied to the art, working habits, and lives of the 38 historical artists selected by the panel of experts as having an old-age style and 68 who did not. How do Cézanne, Constable, Degas, DeKooning, Goya, and the other 33 artists in the former group fare compared to the 68 who did not have an old-age style (Albers, Bingham, Cole, Giotto, Gorky, and the rest)? If samples of 12 or 106 artists (38 with and 68 without an old-age style) are too daunting, a smaller group is available. They are the four artists that the panel of art historians and the three published authorities agreed on as having an old-age style (Cézanne, Goya, Michelangelo, and Titian) and the three that did not (Chardin, Hartley, and Manet).

The set of 20 potentially distinguishing characteristics of the old-age style might also be profitably applied to the artists who received mixed views from four sets of experts (Corot, Degas, El Greco, Homer, Marin, Pissaro, Renoir, and Rubens) to resolve their status. Similarly, the 20 dimensions might be directed to the 29 artists who narrowly missed (by one vote) being selected by the panel of art historians as having an old-age style (Benton, Bonnard, and so on). Some of these artists, after their analysis on 20 dimensions, might move under the rubric of the old-age style.

The 20 possible ways of differentiating artists with and without an old-age style are a useful addition to research and scholarship on the old-age style. I suspect, though, that art historians will disagree. After all, they have disagreed with one another in the past, and will most likely continue to do so. Simonton (1984c) noted that Beethoven's favorite symphonies, the Third and Ninth, are unpopular with audiences today; they are infrequently performed in concert halls. But two piano pieces he "disdained," the "Moonlight Sonata" and "Fur Elise," are "incredibly overplayed" (p. 98).

Nonetheless, whatever the difficulties, research continues. In the next chapter, art historians, laypersons, and artists again contributed to the study of the old-age style. This time their focus was not on identifying artists but on the second critical issue: describing the distinguishing features of works in the old-age style.

CHAPTER 13

Describing Paintings in the Old-Age Style

There is not, I think, a single example of a great painter—or sculptor—whose work has not gained in profundity and originally as he grew older. Bellini, Michelangelo, Titian, Tintoretto, Poussin, Rembrandt, Goya, Turner, Degas, Cézanne, Monet, Matisse, Braque, all produced some of their very greatest works when they were over sixty-five. It is as though a lifetime is needed to master the medium, and only when that mastery has been achieved can an artist be simply himself, revealing the true nature of his imagination.

John Berger, *Success and Failure of Picasso*, 1965

Artists with an old-age style were identified with some certainty (Chapter 12), thereby answering the *identification* question, the first of the two issues basic to subsequent inquiries into this phenomenon of late-life creativity (Chapter 11). The second crucial question is *describing* works done in the old-age style, which is also another way of identifying artists whose paintings show this sign of late-life creativity.

Unfortunately, experts disagree as much on describing works with an old-age style as they do in identifying its artists. Some argue that the key features are related to the work, and they emphasize technique: the amount of detail, the roughness of the brushstroke, and the way lighting is used. Others also concentrate on a work but focus on its mood or feeling: restfulness or turbulence, optimism or pessimism. If not technique or mood, descriptions focus on subject matter: the ages of the people depicted (old or young), a propensity for dignified self-portraits or lascivious nudes, and their treatment (favorable or unfavorable). Some authorities argue that the old-age style is really about the artist, especially their disabilities, and they search for signs of poor eyesight or diminished motor skills in a work. Alternately, it is the artist's feelings about old age or a preoccupation with death. Still other experts look to viewers' reactions: puzzlement or shock, amusement or annoyance, disappointment or appreciation. Then again, perhaps the old-age style is manifested by a large canvas that compensates for a trembling hand. Finally, some writers look for the ways in which late works show the influence of teachers, other artists, or most broadly, the times in which late works were produced.

Some consensus would obviously be welcome. A standardized set of terms could determine if the works of artists believed to have an old-age style (Chapter 12) actually display the appropriate characteristics, and could also clarify the status of artists whose works in the old-age style are unrecognized, uncertain, or questioned. Knowing what to look for in paintings would also guide a search for parallels in exemplars from the other visual arts. Is there a kind of roughness of execution or a disinterest in detail in the late works of sculptors, filmmakers, and architects? Similarly, a checklist of descriptive terms would facilitate a search for similar qualities in the non-visual arts. Do composers, authors, and dancers express something akin to painters' old-age style? The availability of a reliable set of descriptors across the arts could also serve as a template for examining late-life changes in the productions of non-artists. Does the work of aging scientists and scholars reflect a holistic approach to research and scholarship? More generally, ways of doing things that reflect an old-age style might be applicable to the leisure activities of the elderly (as more active or contemplative, familiar or atypical).

As a start toward accomplishing these goals, this chapter presents several studies on descriptors that might be applicable to the old-age style in paintings. As in the previous chapter, participants in the research were art historians, laypersons, and contemporary artists.

ART HISTORIANS' DESCRIPTIONS OF THE
OLD-AGE STYLE

A wide array of descriptions of paintings in the old-age style by experts already exists, albeit with little or no consensus. To determine the best descriptors, and reduce their number, relevant narratives were examined in several published sources on the old-age style in art history (e.g., *Man and his years*, 1954; Munsterberg, 1983; Rosenthal, 1954), gerontology and the psychology of aging (e.g., Achenbaum & Kusnerz, 1978, 1982; Clayton & Birren, 1980). After identifying and extracting relevant sections, the material was edited into single words or short phrases, supplemented by antonyms, and formatted into a checklist. The final checklist contained 79 descriptors: 52 (66%) referred to old age and the balance (27) to youth. The majority (66%, coincidentally) referred to the work: paintings in the old-age style are "rough" and "strong" while those without this feature are "complete" and "technically skilled"; "skill and effort are concealed, not obvious" in an older work while a younger effort "follows rules." Specific technical features pertained mainly to the brushstroke: "bold" in older works and "finicky" in youthful art. ("Bold," "economical," and "simple" described both the brushstroke and the work as a whole.) The terms that remained, about a third (34%), referred to the artist. Older works reveal the artist's "boredom" and "irritation" while younger terms see the artist as "joyful" and "optimistic." A snapshot of the professional literature on old art, elderly artists, the old-age style, aging, and the behavior and accomplishments of the old therefore focus more on art (the work, the product) than on the artist (the person and feelings) and behavior (actions).

The list of 79 terms was shown to six professors of art history with many years of teaching experience (Mean=21.33, Median=26.00); four had previously identified artists with an old-age style (Chapter 12). They selected terms applicable to the old-age style. A descriptor chosen by three or more (50%) of the experts was set as the criterion for an acceptable level of consensus; those chosen by two or fewer were not considered applicable descriptors of the old-age style.

The number of terms chosen by each expert varied widely, ranging from 16 to 59. No term was unanimously chosen, thereby mirroring the wide-ranging scope of the literature. Nonetheless, surprisingly few "uncertain" choices were made (11% of the total; Mean per expert=3.67, Median=1.50); five of the six judges were uncertain in two or less instances. However much the experts disagree with one another, they chose carefully and with confidence.

Thirty-two of the seventy-nine terms (41%) met the criterion as descriptors of the old-age style. Reassuringly, 48 of the choices (61%) were appropriate: 27 which related to old age were accepted and 21 that pertained to youth were rejected. Paralleling the literature, most of the selected terms (84%) referred to the work rather than the artist.

Terms that described paintings in the old-age style (and the number of judges who chose them) were the following: intense, economical, thick, and freely executed (selected by five judges, the highest number); bold, rough, spontaneous, suggestive, skill and effort not obvious, color has a special quality, technique is impatient, expresses profound thought, and *reflects the preoccupations of old age* (four; emphasis added); simple, serene, understated, medium is treated with familiarity, execution is rugged, rough, economical, pursues own ideas, (artist) is an independent person, and has an active mind (three).

The following terms related to youth were appropriately *not* chosen: refined, stylistic, skilled, and finicky (two judges); technically proficient, precise, realistic, objective, very detailed, well formed, composed, follows rules, pictorially structured, even, joyful, and optimistic (one); and clear, emphasizes the appearance of things, happy, and impersonal (zero).

However, 26 terms in the literature related to the old-age style, old, aging, and the like were *not* chosen: simple, raw, stiff, cautious, pessimistic, melancholic, shows solitude, detached, uncompromising, lonely, reverent, a person with fresh ideas, imitative, inventive mind, and *indicates the physical deterioration of aging* (chosen by two judges; emphasis added); primitive, remote, deliberate, bored, emphasizes ideas, irritated, and positive (one): depressed, angry, mistrustful, and remote (zero). Five youthful descriptions of works were inappropriately chosen as applicable to the old-age style: restrained, complex, and not fussy (by four judges); and sophisticated and complete (three).

In Summary. The largest number of art historians (five, or 83%) described paintings in the old-age style in four ways, all of which referred to the work: intense, economical, freely executed, and a thick brushstroke. Appropriately, too, four terms descriptive of young works were *not* selected by a single judge: clear, emphasizes the appearance of things, happy, and very impersonal. Interestingly, the old-age style as "reflecting the preoccupations of old age" met the criterion (selected by four judges) but "indicates the physical deterioration of aging" did not (it was chosen by two judges). The majority of age-related terms were correctly chosen (as old descriptors) or rejected (as youthful referents), over a third (39%) were neither appropriately selected nor rejected. That is, some youthful terms were accepted and older ones were not. The published literature is therefore both right and wrong. Art historians are correct in what they say about the old-age style—but often also frequently incorrect. The problem, of course, is knowing when they are one or the other. This study provides an answer. It also casts doubts on many of the terms considered appropriate descriptions of the old-age style, old-age art, and late art, and these should be questioned if not discarded.

The 32 empirically established descriptors of the old-age style might be validated by applying them to the late works of the 38 artists chosen by art historians as having an old-age style and the 68 who did not (Chapter 12); the two sets should be descriptively distinguished. A more manageable sample is the works by the 15 artists whose prototypical pairs were judged by laypersons as having an old-age style and the nine who did not. Thus, the older works by Eakins, an artist whose pair met the criterion for having an old-age style, should look sketchier, less detailed, have a rougher brushstroke, and so on, compared to his younger effort. However, the younger and older works of Tiepolo, whose works were not judged to have an old-age style, should not be distinguished on the 32 dimensions.

The 32 descriptors can also be applied to artists whose old-age style is in doubt or controversial, and to older contemporary artists whose late-life changes are debatable. In addition, the checklist could also guide a search through personal statements about an artist's work in journals, notebooks, and reviews. Thus, the works of Eakins but not Tiepolo would be described as "intense, economical, thick, and freely executed," and "reflects the preoccupation of old age."

The number of terms (32) is sufficiently large and diverse to allay the fears, of art historians and other scholars in the humanists of losing the individuality of art, ignoring the exceptional, and by-passing the unique. For scientists, the 32 descriptors can be a basis for formulating hypotheses about the kinds of cognitive, personal, developmental, and motivational traits that persist into old age. For example: physical and sensory declines with old age were emphasized in the professional literature, but not among the judges who failed to select "indicates the physical deterioration of aging" as an apt descriptor. Thus, old age, at least in the creative domain of painting, is not, once again, a time of decline. The challenge for empirical inquiry is to determine the areas in which losses in strength, hearing, and sight do play a role, since it is evidently not in artistic creativity, and moreover, establish the ways in which artists compensate for these handicaps.

Before these advanced studies proceed, though, other ways of describing the old-age style should be explored. To this end, laypersons' contribution to describing the old-age style is examined next.

JUDGMENTS OF YOUNG–OLD PAIRS BY LAYPERSONS

In the previous chapter, laypersons distinguished 15 pairs of young–old art, and these pointed to artists with an old-age style, and the 9 artists who did not show a difference. Four judgments differentiated the pairs: differences in style and the amount of detail, authorship (the members of the pair appeared to be painted by different artists), and age-of-production (one member seemed to be by an older artist). Comparisons between the effectiveness of these types of judgments suggest their usefulness as descriptors.

Of the four ways of judging pairs of paintings, authorship was most effective: 65% of the judges indicated that older members of the pair looked as if they had been done by another artist; the percentage increased to 76% when eight pairs by different artists were included. Somewhat less discriminating were judgments of style: 58% of the judges reported a difference between the younger and older pairs, a percentage that increased to 63% when judgments of the different-artist pairs were included. The remaining two kinds of judgments also distinguished the pairs, at about the same rate as the above—but in unexpected ways.

The pairs' age-of-production was *incorrectly* judged by a majority of judges (62%). That is, older works were wrongly judged to be painted by younger artists. The contrariness of these judgments was explained by the judges' written comments in a post-study question-naire. Older artists, they reasonably assumed, have greater training, skill, and expertise than younger and less experienced artists. The judges therefore believed that older art should display more technical virtuosity and skill than youthful examples, given aging artists' greater maturity and competence. But older art with an old-age style, accord-ing to art historians, is formless, ambiguous, fragmented, unfinished, and hurriedly done. Thus, untrained judges erroneously interpreted the more dazzling youthful works as the product of older artists, indirectly affirming experts' expectations.

Judgments about detail were also anomalous: Contrary to art his-torians' expectations, once again, older works were judged to have more detail by 61% of the judges. These unexpected judgments can be explained on methodological grounds. Unlike the other tasks, detail was judged in a non-laboratory setting, under full illumination, mak-ing it difficult to pick out details in the slides that were photographed from small black-and-white reproductions. Consequently, the amount of detail within and between people, places, and things was difficult to detect. (The reason why older but not younger works were judged to have more detail hinges on other unusual circumstances of the detail task, compared to the other judgments; see the discussion in the origi-nal article [Lindauer, 1993a].) It is therefore difficult to interpret the usefulness of the paradoxical results for judgments about detail.

The four judgments, though, cannot be applied indiscriminately to young and old works, since their effectiveness varied with the type of art and the artist. Consider judgments about authorship. Abstract art was differentiated by more judges than figurative and non-figurative repre-sentational art (74%, 60%, and 62%, respectively). Authorship's effec-tiveness also varied for specific artists. It clearly differentiated the pairs by the American Eakins (the pair is identified in Chapter 12) and the British Reynolds' (1723–1792) *Portrait of Miss Henriett Edgcumbe* (1756) and *Portrait of Sophia, Lady St. Asaph With Her Son* (1787): 87% and 81% of the judges, respectively, considered the older example to be painted by an artist other than the one who did the younger work. (Paintings previously described by titles, artists' birth and death dates, and nationalities [Chapter 12] are not specified again.) On the other hand, only about a third (32%) of the judges judged the pair by the American Cole (1801–1848), *Landscape: Scene from the Last of the Mohicans* (1827) and *Mount Aestna from Taormina, Sicily* (1844), to be painted by different artists.

Style judgments also varied with the type of art. Older figurative works differed more in style than older non-figurative works, according to 68% and 54% of the judges, respectively; and both types of art differed more than pairs of abstract art, which were as often seen as similar in style to one another as dissimilar (by 50% of the judges in each case). Style differences for some artists differed from the general trend, too, just as authorship did. For example, few judges (12–17%) found a style difference between the younger and older non-figurative pairs of three artists: the Italian Guardi (1712–1793), whose pairs were *Venice: Church of Santa Maria Della Saute* (c. 1750) and *Grand Canal with the Church of Santa Mariate* (1780s); the American Copley (1737/38–1815), represented by *Portrait of John Bours* (1758–1761) and *Portrait of Sir Robert Graham* (1804); and the abstract pair by Hoffman (pairs were named in Chapter 12).

Judgments about age-of-production, like authorship and style, varied with the type of art and artist. Thus, 81% of the judges erroneously attributed older abstract works to younger artists, while 64% of the judges erred for figurative works but hardly differed on non-figurative works (52%). Among specific artists, the older figurative art of Eakins and Reynolds were correctly chosen as older by a slim majority (by 54% and 52% of the judges, respectively), but only 18–29% so judged the non-figurative pairs by the German Kirchner (1880–1938), whose pairs were *Flower Beds in the Dresden Gardens* (c. 1910) and *Mountain Forest (Forest Path in Summer)* (1927–28), and the American Innes (1825–1894), who painted *The Sun Shower* (1847) and *Near the Village, October* (1892). The older work by the American Tobey (1890–1976), represented by the pair *Broadway* (1935–36) and *Ghost Town* (1965), was more often correctly identified than any other abstract artist (49%), while the abstract works by the American Klein (1910–1962), *Chatham Square* (1948) and *Palladio* (1961), were hardly ever correctly judged (by only 7% of the judges).

Detail differences were greatest for figurative pairs of art (according to 64% of the judges) but were about evenly divided for pairs of abstract art (51%). (There were too few non-figurative examples in this task to make comparisons meaningful.) All the judges, furthermore, judged the older abstract work by Klein as having more detail than his younger work, but not a single judge perceived a difference in detail between the young–old examples of Klee and the Spaniard Picasso's (1881–1973) *Nude* (1909) and *The Painter and His Model in the Garden* (1963). Relatively few judges (33–44%), furthermore, saw detailed differences between the younger and older figurative works by Eakins, or the pair by the American Sargent (1856–1925), *The Oyster Gatherers of Cancale* (1878) and *Two Girls Fishing* (1912).

Summing Up. The most effective way of distinguishing young–old pairs of art with an old-age was their authorship. That is, if an artist's young and old works seem to be painted by a different artist, then the latter was probably painted in the old-age style. Judgments about style differences were slightly less effective. Modestly useful was the apparent age-of-production of an artist's work, albeit quixotically: Old works looked as if they were painted by *younger* artists. Thus, works done in the old-age style would *not* look as if they were painted by old artists, at least by untrained observers. Most puzzling, though, is the finding for detail differences: Older works looked as if they had *more* detail than younger ones, contrary to art historians' expectations. The reliability of this distinction is problematic, though, since it probably reflected the less than optimum conditions under which the pairs were viewed.

The effectiveness of each of these four ways of judging paintings, moreover, depended on the type of art, whether it was abstract or representational, and if the latter, figurative or non-figurative. The usefulness of the judgments was also linked to the specific artist whose works were judged. Consequently, the four types of judgments should be applied judiciously, taking into account the type of art and artist. Other kinds of judgments are possible, of course: complexity, meaningfulness, and interestingness, to name just a few. Of the many possible, preference was investigated next: If one member of a young–old pair was liked or disliked more than the other, this choice may hint at an artist with an old-age style.

Preferences for Young–Old Art. Laypersons are often criticized for reacting to art in terms of "liking," which seems to be a rather superficial reaction to art. However, an initial emotional impression sets the stage for subsequent and more subtle reactions. Thus, a work done in the old-age style might be liked more (or less) than an example without this feature. To investigate the usefulness of preference as a possible indicator of the old-age style, a group of lay undergraduates indicated which member of the 24 young–old prototypical pairs was liked more. (These were the same pairs previously shown.)

Most judges (60%) preferred younger over older works, a percentage that increased to 70% when non-representational works were excluded. The latter was disliked more than figurative and non-figurative art, which were about equally liked, again indicating that judgments depend on the type of art. Preferences also varied with the artist. Although abstract art was generally not liked, the older work by Klein was an exception: it was liked by nearly half of the judges (44%). Among other kinds of older art, which were generally liked, there were two exceptions: the figurative pairs by the German Corinth (1858–1928) and Eakins (6% and 16% of the judges liked them, respectively). (Corinth was represented by *Portrait of Franz Heinrich Corinth, the Artist's Father* [1888] and *The Black Hussar* [1917].) Disliked, too, contrary to the overall trend for non-figurative art, was the older example by the American Bellows (1882–1925), represented by *Summer Night—Riverside Drive* (1909) and *The Picnic* (1924); it was liked by only 11% of the judges. Also disliked was the older work by the Frenchman Pissaro (1830–1903), represented by the pair *Path By the River (Near La Varenne—St. Hiliare)* (1864) and *Varengeville, Auberge du Manoir, Gray day* (1899); only 21% of the judges preferred the younger example.

Undergraduates may prefer younger art because it matches their youthful exuberance, joyfulness, and playfulness, while the serenity, calmness, and tranquillity of older art does not; the subdued mood of late works may be more congruent with older viewers' sensibilities. If these assumptions are correct, it may explain why art historians and critics dislike and reject works done in an old-age style by artists older than them. Thus, differences in age may influence preferences for the old-age style, a possibility that awaits future research.

But at least for untrained young judges, preferences may be another way of distinguishing young–old pairs with an old-age style: an older work is disliked more than a younger one. Whether this holds true for trained older viewers remains to be seen. The relationship between age and art is examined in Part V; for a brief discussion of the role of experience with art in judging paintings, see Lindauer (1970a).

Old age and experience with art were combined in the 60- to 80-year-old artists whose answers to a questionnaire were reported in previous chapters. The following report presents their replies to a questionnaire item on the affects of physical and sensory losses on their late works. Signs of these kinds of handicaps in late works may be another indicator of the old-age style.

THE ART OF OLD AGE AS A REFLECTION OF PHYSICAL AND SENSORY LOSSES THAT MAY ACCOUNT FOR THE OLD-AGE STYLE

Old age inevitably leads to declines in health, strength, motor abilities, hearing, and sight. These losses are especially critical for painters who depend on their ability to perceive color, form, and space, on motoric coordination in handing a brush and mixing paint, and on strength and mobility in handling an easel and working outdoors. Their debilities may explain declines in creativity with age (Chapter 3) as well as the old-age style (Chapter 11). If these losses are found in art, they would be another useful way of describing works in the old-age style. Frailty, poor eye–hand coordination, and failing eyesight could explain the reduced detail, rougher brushstroke, and the formlessness of paintings that characterize the old-age style.

These possibilities, however, were not apparent in the reports of creative artists in their 60s to 80s on changes in their work with age (Chapters 9 and 10). Physical and sensory losses ranked third in salience, and moreover, their primary influence was on the quantity (productivity) of late-life art and not its quality, the general approach to their work, and the sources of ideas. When poor eyesight, declining health, and lowered energy were mentioned, it was by artists in their 70s and 80s rather than in their 60s, the age at which the old-age style is supposed to emerge, at least among historical figures. For the relatively few artists who reported health-related problems, losses were compensated for, adjusted to, or worked around. For example, as artists became frailer, they shifted to an easier medium, reduced the amount of time they worked, varied the time of day at which they painted, or changed the location of their work. As artists became more infirm, they painted at home, relied more on memory or used photographs instead of going out into the field, and they depended on others for help, for example, in carrying an easel. These comments, however, were made spontaneously in response to general questions about creativity. Whether these artists' work had been affected by losses in sight, for example, was not directly asked.

Fortunately, another part of the questionnaire did, and the results are presented here. The artists were explicitly asked about the impact of physical and sensory losses on their work as they aged. Although the question was not directed at the old-age style (for reasons explained earlier, e.g., the terms' negative connotations), the answers may indirectly bear on descriptions of such works. Thus, if aging's handicaps do not impinge on the work of aging artists, then they are not likely to show up in the paintings of artists with an old-age style.

As it turned out, most of the 87 artists (60%) who were questioned reported that physical and sensory handicaps did not impair their work; only a third (33%) wrote that their art had been affected by these losses. (The replies of a small percentage [7%] were unclear or mixed.)

> Even though I am quite deaf in one ear and totally deaf in the other (I have two hearings aids), I spend more time now (and have the past four years) at my easel and drawing board. I spend from 10 to 14 hours alone in my studio except for lunch and dinner and an hour or two watching television. (A 73-year-old man)

> My vision is still 20/20 and I may have a cataract started. I can stand up and even kneel down a little better than I could in the early 1980s. My hands are still OK. My memory is not as good as it used to be. But I think my standards are unaltered. (An 83-year-old man)

When loses were reported, they were more often sensory than physical (65% and 22%, respectively; unclear and mixed responses are not included). Of the sensory losses, more were visual than auditory (58% and 42%, respectively). Compensations were the rule, though. "I can't see as well as when I was young," reported a 62-year-old man, "but I have glasses and magnifiers so I feel I can control the mechanical part of producing my art works."

Most physical problems (77%) referred to reduced strength and energy rather than restricted or impaired movement. These losses, like sensory ones, were compensated for. "So far," wrote a 72-year-old woman, "I have only slightly diminished energy to contend with—but that is no problem. I just work for shorter lengths of time and in the morning when my energy is highest." A 76-year-old man was more specific on the way he coped.

> I spent years working on location, outdoors, and using my car. My physical limitations make that way less practical. Now I use a camera and quick sketches on location and do most of my painting in the studio.

In short, whatever the types of physical or sensory infirmities artists overcame them. "I tire a bit earlier in the day, " said a 71-year-old man. "I recognize that and don't concern myself with it." Other compensatory strategies are illustrated in the following comments by three women, 68-, 77-, and 87-years old, respectively.

> I miss the kind of reluctant help my late husband could give in making presentation 'props' and transportation. This is not a devastating loss,

though. I simply transfer or emphasize art activities that require less of the above. As an artist, I may be better off than before, with complete independence.

I can no longer make very large projects, but making small things can be rewarding also. My energy has diminished somewhat, and a lot of time has been lost recovering from surgery, but I have never stopped working. I have a compulsion to make things of my own design. I am fortunate in that my mind seems to be intact.

So far, no loss due to age has evidenced itself, although it's more difficult to carry heavy pieces, especially for exhibitions and displays. On the positive side, I am more selective due to years of experience. My back tires, hence I spend less hours at the Bench.

For another 66-year-old woman, overcoming physical adversities made her a better artist.

I wouldn't object to a slow up in quantity because of my heart-brain problem [she had an unspecified injury or surgery] but I will not be able to live through a stoppage of my quality. My whole later life has been a concerted effort of body, history, heart, and experience. No aging process has affected me, nor will. I have always been hard of hearing or deaf, so it is hard to say if there is an effect. An artist is forced to be isolated. Since it is difficult for me to work with people, I am in forced isolation to a great extent. Getting deafer has improved my art but maybe not me.

For some artists, having a disability was less important than their attitude toward it. "My eyes are getting worse but that's just an excuse," wrote a 66-year-old man. Declines were also seen as relative, cutting down on what was still a rather heavy work load, as this comment from a 69-year-old woman indicated. "I can no longer work a 15 hour day."

In Summary. Most of the contemporary artists in their 60s to 80s indicated that the physical and sensory debilities of old age did not seriously impair their creative efforts. These explicit replies to a direct question about handicaps corroborated the spontaneous comments made earlier to broader questions about creativity, where physical and sensory factors ranked third in importance and affected how much was painted rather than how well they were done.

Elderly artists of repute overcame aging's inevitable physical and sensory limitations, at least with respect to doing art. The art of old age is ageless, in the sense that it is not impoverished by aging's inevitable handicaps. Decreased energy, poor eyesight, losses in eye–hand coordination, and other debilities did not seriously detract from the art of old age. Instead, artists made practical adjustment to aging's debilities rather than surrendered to them. Physical and sensory impairments, though, may be handicaps outside of the artists' studio, in everyday life, in shopping, traveling, and engaging in social activities. There may of course be other impairments that were not asked about or mentioned, such as losses in memory and attention.

If declining health, physical limitations, and sensory restrictions did not detract from aging artists' late-life creative achievements, then there is little reason to expect them to influence the old-age style or to find signs of these losses in their work. Aging's debilities, at least of the kind investigated here, are likely to inconsequential, secondary, or minor determinants of the old-age style, as they were in the case of late-life creativity in general.

The *absence* of signs of physical and sensory handicaps in paintings can therefore be added to the list of 32 terms and 5 kinds of judgments that described the old-age style. Obviously, there are many other ways of describing late-life paintings, as the literature so profusely indicated, and some may be, upon investigation, germane to the old-age style. A few possibilities are examined next.

OTHER DESCRIPTORS OF THE OLD-AGE STYLE

Young and old art might differ in many ways, but probably most obviously in subject matter. This characteristic and a few other possible ways of distinguishing the old-age style touched on by the research are reviewed next.

Subject Matter

Subject matter differences between young and old paintings by aging artists are easily noted and could be a basis for describing works done in old-age style. Late works might favor landscapes, mirroring aging artist's reflective sensibilities, or they might depict old people, indicative of the aged's preoccupation with growing old. Religious and spiritual themes, for the same reasons, might also dominate as artists approach death. Younger works, in contrast, might emphasize erotic nudes, turbulent cityscapes, and celebratory historical events, paralleling the excited, exuberant, and active temperament of youth.

To investigate these possibilities, age-related differences in subject matter were inspected in the collection of 72 young–old prototypical pairs in the Rosenthal (1954) catalog of the late-life paintings by 70 artists. (In addition to artists already noted in this and the previous chapter, the other pairs were by the following artists: Canaletto, Chagall, Constable, Degas, Delacroix, Dubuffet, El Greco, Fragonard, Gainsborough, Gauguin, Gorky, Greuze, Hals, Homer, Ingres, Kandinsky, Lourrain, Magnasco, Peale, Pollock, Poussin, Rembrandt, Renoir, Rubens, Ruisdael, Strozzi, Toulouse-Lautrec, Van Dyck, Van Gogh, Vigee-Lebrun, Villon, and Vuillard.) The collection of late art included older artists with and without an old-age style, some of whom were questionable or questioned, as well as a handful of artists who died young but had a late-life/old-age style, and others who simply represented the art of old age. If most of these lack certain subjects, it stands to reason that art with an old-age style is not likely to depict them either. For example, if physical and sensory handicaps and aging individuals are (or are) not depicted in nearly 100 examples of various kinds of late-life art, then they are likely (or not) to be found in works done in the old-age style. Thus, the presence of certain subjects in old age art, or their absence, could be another set of descriptors for works painted in the old-age style.

Figurative works were the largest category of art in the Rosenthal collection; it accounted for 55 pairs (38% of the total). Included in this group are portraits, genre works ("a servant"), and individuals or groups in conversation, observing, or doing something. Differences between the young–old pairs in the number and type of people and activities, if any, were therefore easily observable. Figurative examples were a good choice for another reason: people were about equally represented in the younger and older works (27 and 28, respectively), thereby simplifying quantitative analysis.

Alas, the art was unrevealing. Only three figurative works depicted an obviously older person, and two were painted in the artists' old age. Women, often as nudes, appeared 21 times, but they were about equally represented in younger and older art (11 and 10 times, respectively). Children were more often seen in younger than older art (7 and 3, respectively), but the numbers are too small to be more than tentative, at best.

The next largest category of art in the Rosenthal collection were 32 landscapes and still-lifes. Scenes of nature may suggest the reflective proclivities of aging artists. Disappointment again. Pastoral vistas and bowls of flowers were distributed fairly equally between younger and older works (18 and 14, respectively). Other types of art, although infrequent, were suggestive. The six cityscapes characterized younger works only, perhaps indicating younger artists' greater vigor and heightened level of activity. Promising, too, were shifts with age from representational to abstract art, which changed more often from young to old art than the reverse (13 and 6, respectively), perhaps indicative of older artists' penchant for formlessness. Religious subjects, though, characterized more of the younger than older works (7 and 3, respectively); one might reasonably expect the opposite as older artists approached death and became more spiritual. Despite some encouraging hints, the various subjects in the prototypical young–old pairs of art in the exemplary Rosenthal collection differed too little to suggest descriptors of an old-age style.

A more subtle feature related to subject matter is the direction of a depicted person's gaze (Bryson, 1983). As artists grow older and perhaps more contemplative, spiritual, and religious, the gaze of individuals in paintings might change, looking up rather than to the side or directly at the viewer. However, like the more obvious kinds of subjects, the direction of the gaze of people in younger and older works did not appreciably differ from one another: 17 people were looking away from the viewer (either left or right) and 10 were looking toward the viewer, and neither direction was distinguished by the art's age.

The next step might be to look at facial expressions in portraits: joy and happiness in youthful works, and frustration and sorrow in later works. Judgments of facial features, though, are less easily described and verified than more explicit subjects, and viewers' interpretations may be idiosyncratic. Subtle and personal, any differences noted would probably evoke contentious and unresolvable disagreements, of which there are already a surplus in the literature on the old-age style.

A better strategy would be to look for subject matter differences in a large collection, say of several hundred young–old pairs. The findings from the small Rosenthal collection of less than a hundred suggest several possible distinctions to look for in a large sample: the number of children portrayed, cityscapes, and religious subjects, and shifts to abstract works. Differences might be further maximized by examining only the works of very old artists who died in their 80s, when the effects of aging might be sharpest.

Canvas Size

A simple and readily observable characteristic of art that might distinguish young and old works is canvas size. This dimension is provided with each painting along with its title, artist, and year signed. Two reasonable and opposing predictions are possible. On the one hand, large canvases might be preferred by aging artists because it gives them more space in which to express characteristics associated with the old-age style, such as a holistic treatment, undifferentiated forms, a rough, bold, and broad brushstroke, and an unconcern with detail. On the other hand, small canvases might be chosen by aging artists because they are easier to handle, thereby conserving waning strength and minimizing motoric limitations.

The 72 young and old pairs in the Rosenthal collection were again examined. Unlike subject matter, canvas size clearly distinguished pairs: Older works were generally larger than younger ones (Mean=1,847 and 1,244 square in., respectively). The difference, furthermore, varied with specific time periods. The greatest size difference between younger and older canvases occurred in two time periods: the 17th century (Mean=1,422 and 2,336, respectively) and the later half of the 20th century (Mean=1,257 and 3,041, respectively). Size difference were less marked, though, in the early half of the 20th century (Mean=1,270 and 1,437, respectively). For the 20th century as a whole, though, canvases of older works were considerably larger than younger ones (Mean=2,239 and 1,264, respectively).

Four artists with an old-age style (Chapter 12) illustrated these differences. The older canvases for Eakins, Tobey, Monet, and Klee as a group were about five times larger than their younger works (Mean=2,548 vs. 466 square in., respectively). Eakins' younger work, though, was marginally larger than his older painting (396 vs. 320, respectively). Now consider the three artists whose pairs failed to show an old-age style: Hoffman, Manet, and Tiepolo. Their older works were also larger than their younger ones (Mean=2,571 vs. 1,129, respectively; Manet was an exception: his younger work was considerably larger than his older work: Mean=1,178 vs. 304, respectively). Size differences for artists without an old-age style, though, were much less than those for artists with a late-life change: The former were about two times the size of younger ones, compared to five times for the latter.

Canvas size differences are therefore a better indicator of old-age art, and by inference, the old-age style, than subject matter differences, although their superiority depended somewhat on the time period and the particular artist. Keep in mind, too, that subject matter affects canvas size: historical paintings usually require larger canvases than still-lifes. Changing techniques and fashions also influence canvas size. Some contemporary artists also prefer huge canvases. (Why canvas sizes for older works in the 17th century were so much larger than youthful efforts at that time is puzzling and probably requires a specialist of this period to explain it, although more examples must verify this finding.)

Now that the facts are in, it makes sense for old artists to prefer large surfaces. Having lots of space to work on compensates for losses in eye–hand coordination and failing eyesight. More space also facilitates a looser and rougher brushstroke, encourages broad and holistic treatments, and makes close attention to details unnecessary, all of which are concomitants of old age and characteristics of the old-age style.

Other Characteristics of Late-Art

Other discernible features of art may distinguish works done in the old-age style, such as the artists' nationality and life span. These possibilities were examined for the 15 pairs with an old-age style and the 9 without (Chapter 12). No distinctions along these fundamental lines emerged. For example, the two sets of artists' life spans did not (statistically) differ (Mean=73.85 and 68.54, respectively). Irrelevant, too, was the time period in which the 16 representational works were done (abstract art was excluded because it is exclusively a 20th century phenomenon). The sample is small, though, so it may pay to reexamine nationalities, life spans, and time periods for the 38 and 68 artists with and without an old-age style (Chapter 12).

DESCRIBING PAINTINGS WITH AN OLD-AGE STYLE

Add large canvas size to the empirically determined descriptors that differentiated works done in the old-age style, together with the other distinctions already noted. To review: When art is paired into young–old examples, five kinds of judgments were effective. For example, older works looked as if they were done by a different artist than younger paintings and were less liked. Thirty-two terms were selected by art historians to describe the old-age style. Such works are intense, economical, thick, freely executed, and so on. Add these 38 descriptors (canvas-size, 5 kinds of judgments of pairs, 32 descriptors of works) to the 20 possible descriptions of artists with and without an old-age style (Chapter 12), such as greater knowledge, innovativeness, energy level, maturity, and self-acceptance. There are now more 50 ways (38 for works and 20 for artists) that might describe old art and aging painters with an old-age style. To these 58 add the 5 ways in which young and old art do *not* differ: old works do not show the effects of physical and sensory losses, and young–old pairs do not differ in subject matter, time period worked, nationality, and life span. There are therefore quite a few empirically established ways, 63 to be exact, to describe works and artists with and without an old-age style.

In addition to this taxonomy of 63 dimensions that describe works with an old-age style there are at least 38 artists with an old-age style and a minimum of 68 without one (Chapter 12). These numbers (63 descriptors and at least 106 artists) are large, but not as large as the literature contains. If a smaller and more manageable number of artists is preferred, take the six artists who were judged by both untrained viewers and art historians as having an old-age style (Eakins, Goya, Klee, Mondrian, Monet, and Picasso) and the six who were not (Copley, Hoffman, Manet, Marin, Stuart, and Tiepolo).

Art historians now have two major tools—the names of artists with an old-age style and descriptions of them and their work—with which to examine thousands of old paintings and hundreds of long-lived painters. These include the many not included in the studies reported here, the varieties of modern and post-modern art, so-called primitive and ethnic art, and folk art.

Knowing what artists to look at and what to look for advances other issues, too, such as explaining the old-age style (Chapter 11). Furthermore, the methods that identified painters and described paintings can also guide the identification of artists and descriptions of works with an old-age style in different realms of art, in other professionals, and for aging non-artists in general.

The old-age style is an impressive example of late-life creativity, and more generally, an extraordinary illustration of what the elderly can accomplish. It is also further evidence for the reality of successful aging, encourages an optimistic attitude to old age, and supports a positive perspective on aging. These and other implications of the studies on the old-age style are discussed in the next chapter.

CHAPTER 14

Beyond the Old-Age Style, Old Art, and Aging Artists

> For the past eight years I have started each day in the same manner. It is not a mechanical routine but something essential to my daily life. I go to the piano, and I play two preludes and fugues of Bach. I cannot think of doing otherwise. It is a sort of benediction on the house. But that is not its only meaning to me. It is a rediscovery of the world of which I have the joy of being a part. It fills me with awareness of the wonder of life, and a feeling of the incredible marvel of being a human being.
>
> Pablo Casals, *Joys and Sorrows*, written at age 93.

A number of facts were learned about the old-age style in paintings: Its artists, their number (Chapter 12), and the distinctive features of their work (Chapter 13). These major accomplishments can clarify questions that have been heatedly debated in art history with more contention than consensus. With two reliable starting points—which artists to look at and what to look for—research can move forward confidently on other issues. An explanation of the old-age style, for example, is likely to be rather straightforward rather than buried in mysterious, grand, and indecipherable forces deep within the psyche. Probable reasons are likely to revolve about artists' knowledge and self-acceptance, their ability to make practical accommodations to growing old, and intelligent compensations for physical and sensory losses. To test these assumptions, up to 63 empirically determined characteristics of the old-age style (Chapter 13) can be applied to at least 38 historical artists with an old-age style and 68 without one (Chapter 12). The task can be simplifying by reducing the number of artists to the 15 whose prototypical young–old pairs clearly showed an old-age style (Eakins, Manet, Kirchner, Mondrian, Guardi, Klee, Bellows, Cole, Goya, Innes, Picasso, Pissaro, Reynolds, Sargent, and Tobey) and the 9 who did not (Marin, Tiepolo, Hoffman, Klein, Corinth, Copley, Leger, Manet, and Stuart); and the number of descriptors can also be shortened to include the four major characteristics of works done in the old-age style (intense, economical, freely executed, and a thick brushstroke) or the five most important reasons reported by contemporary artists for changes in creativity (increased knowledge, greater self-acceptance, physical limitations, having more time, and new circumstances). In pursuing these and other questions about the artists' old-age style, artists with and *without* this style have to be compared. Did aging artists with an old-age style suffer poorer eyesight, limited mobility, and unsteady eye–hand coordination compared to those who did not change their style? Do artists with and without an old-age style differ in their independence or courage (however these terms are defined)?

Research in Part IV also raised new questions not posed in the introduction (Chapter 11), such as the influence of viewers' age and experience with art on judging the old-age style. Scientific inquiry, by its very openness, poses questions of its own making that did not exist before research was conducted. These touch on wider issues that go beyond art and artists, discussed next.

CREATIVITY

The old-age style explains creativity that is redirected as well as enhanced, why it apparently declined and ended for some artists, was restored and (re)discovered with time for others, as well as why certain works were misinterpreted in their time. Marked changes in the works of aged artists were mistakenly judged as artistic failures or flawed because of a personal breakdown but were actually signs of an old-age style that were unappreciated and unrecognized. The old-age style, moreover, is further proof that creativity continues beyond the middle years, into the 60s and later, and that new forms of creativity can emerge late in life. The extension of both the span of creative expression and its forms counters the prevailing opinion, in professional circles and popular thought, that creativity reaches its maximum in youth, lasts but for a brief moment in time, ends shortly in early middle-age, and thereafter precipitously plunges, disappearing by old age (Chapter 3). Doubts, too, are cast on the predictability and consistency of creativity once its path is set in youth. Moreover, the old-age style accounts for secondary peaks in creativity (Chapter 8) and its late-blooming creativity ("Grandma Moses"). The old-age style also explains the extraordinary achievements of some old people in a wide range of occupations (Chapter 1). These possibilities for late-life growth support programs for the aged that have the potential of creative renewal.

ART

If the creativity of old painters can take new directions, so too can aging composers, writers, sculptors, filmmakers, architects, and other kinds of artists. To test this possibility, the methods used to study old painters and late-life paintings can be applied to other kinds of artists and works. The names of renowned old artists in music, literature, and other fields are readily obtained from standard references, as painters were. Alternatively, musicologists, literary critics, and other experts in different areas of the arts can be polled, as art historians were, on composers, writers, and others whose later works drastically differed from younger efforts. In addition, pairs of young and old exemplary works by sculptors, architects, and others kinds of artists can be judged for their identify of authorship and the like, as was done for prototypical pairs of paintings.

The critical features of the old-age style in various forms of the arts can be also be uncovered by appropriately editing the 63 descriptors for painters and paintings: late sculpture as "rougher," architecture as "bolder," music as "more intense," and so on. Alternately, a new list of descriptors can be developed from the specialized literature in the different fields of the arts, and their adequacy checked by surveying experts in these fields, in the same way paintings were. Aging sculptors, architects, musicians, and other artists can also complete questionnaires, as painters did, on changes in their work as they aged, on whether their late-life work differed from earlier efforts, and if so, how and why, and the extent to which physical and sensory losses affected their work. Studies across the arts can be compared in order to disclose parallels in late-life creativity and the extent to which the old-age style is universal.

The findings for the old-age style in painters and paintings (Chapters 12 and 13) can also be a basis for interviewing contemporary painters to determine if this kind of late-life creativity occurs today, and to the extent it does, if it differs from historical artists. Is it the case that new kinds of art and materials, improved working conditions and financial resources, and living longer and healthier affect the appearance if not the existence of the old-age style? If this change does occur, it may emerge later than 60, perhaps as late as 80, and the phenomenon would be re-named the "*very* old-age style." Critics of aging contemporary artists, once sensitized to the revised nature of the old-age style, will become more cautious in evaluating unexpected changes in late-life works. "Poor art," like apparent losses in creativity, would be reconsidered as manifestations of the old-age style.

A large number of factors describe the old-age style, and these depend on the type of art, the specific artist, and even the time period in the case of canvas size (Chapter 13). Thus, evaluations of the old-age style have to look for many factors and their combinations. The failure to do so explains why art experts differ so much from one another. Each authority concentrates on one or a few features of the old-age style, such as the amount of detail, a shaky brushstroke, the composition as a whole, the mood of the artist, signs of visual losses, and the like. Studies of youthful and later pairs of paintings (Chapters 12 and 13) illustrate the consequences of taking too narrow a perspective on the old-age style. Consider the age-of-production task, where each member of a pair was judged on whether it looked as if were done by an older artist. On this basis alone, the older works by Cole, Copley, Eakins, Guardi, Monet, Reynolds, and Tobey would have been labeled as painted in the old-age style. However, when other judgments were included, some of these artists maintained their distinctiveness, others did not, and different artists came to the fore. Thus, the special qualities of the older works by Eakins, Tobey, and Monet held up across more measures than the ones by Cole, Guardi, and Reynolds, while Copley's young–old pair was undistinguished on subsequent measures. Only Eakins' older work continued to stand out from his younger effort when four measures were taken, Monet and Tobey did so on three, two instances for Cole, Guardi, and Reynolds, and only one for Copley. Had distinctions between artists been based on just one dimension, then different artists with an old-age style would have been identified.

Discussions by experts of the old-age style would become more consistent if they depended on an agreed upon set of descriptors. But with more than 60 ways of distinguishing the old-age style, a shorter set would be less cumbersome. Good candidates are the five dimensions that distinguished young–old pairs: similarity of style, identity of authorship, apparent age of artist, the amount of detail, and preference. However, given the variety and individuality of artists and paintings, historians would probably want to rely on other distinctions. If the 63 applicable to paintings are insufficient to describe the nature of various art forms, the methods of the studies reported here can be modified. Hence the value of providing details of the research.

Whatever the number and type of descriptors, the task should not be too difficult. After all, undergraduates untrained in art selected works by artists with and without an old-age style, and did so with considerable unanimity (Chapter 12). More remarkable, their choices substantially agreed with the experts (Chapter 13) despite major differences between the two groups' backgrounds, experience, and training as well as the methodology of the studies in which they participated: art historians based their judgments on the reputations of artists and undergraduates compared 24 pairs of young–old works on several tasks. The correspondence between undergraduates and experts is unexpected in light of the contentiousness in the published literature, where art historians rarely concur with one another (Chapter 11). The reason the research participants agreed with one another, and published accounts do not, is because the former groups were given detailed instructions that were carried out under controlled conditions. In the "real world," experts make their judgments in largely unknown ways with criteria that are unspecified.

The ability of laypersons to pick up the subtle distinctions of the old-age style is good news to museum directors and art educators whose visitors and classes are made up of a public that also has little background in art. Despite the skepticism of elitists, the untrained observers' sensitivity to the nuances of art should be encouraging news to publishers of art books and art reproductions. Great art, as is sometimes claimed, is capable of reaching the "uneducated eye." Youth and inexperience were not fatal handicaps in responding to art.

LATE-LIFE DEVELOPMENT

There are good reasons to believe that an old-age style exists in aging scientists, scholars, and old people in general. Why should changes in late-life creativity be limited to painters or artists only? People age in similar ways whether they are eminent, talented, or ordinary; aging is universal. Openness to new learning, a primary factor in the old-age style as well as late-life creativity, applies to aging professionals in all fields and to non-artists. Aging artists, elderly experts, and senior citizens differ from each other in education, training, temperament, experience, and other ways. But they also age in similar ways, and this should include the potential for an old-age style. This change might be manifested in various ways: a broader and less-detailed perspective in thinking, work, and other activities late in life; an increase in holistic thinking, compared to narrower modes that dominated the younger years; and activities that unexpectedly differ from earlier ones and from aging colleagues without an old-age style.

It should be relatively easy to find individuals who, late in life, sharply changed their goals or means to accomplish them. There is a large pool of aging, active, and competent scholars and scientists

(Birren, 1990/Special issue; Birren & Deutchman, 1990; Munnichs, 1990; Shimamura et al., 1995). Signs of an old-age style might be shown by redirecting their earlier work, becoming less specialized or more critical, giving up long-term conceptions, welcoming if not seeking new developments in their fields, and moving beyond detailed inquiries that merely added footnotes to younger accomplishments. Indications of an old-age style might also be expressed by taking time out from work in order to reflect on its larger meaning, pursuing new problems, and doing so in unusual ways. For the gifted who reached the ages of 70 and 80, old age was "a special time of fulfillment, independence, increased opportunity for selfhood, and freedom and release" (Schneidman, 1989, p. 685).

Senior citizens might indicate a nascent old-age style by radically changing their thinking and behavior after retirement in ways that demonstrate mental agility and behavioral flexibility, while their elderly compatriots remain unchanged, limited, or dimmed. They might take a fresh look at themselves, change their lives and relationships, go back to school, or start a new career. Or they could develop an intense and dedicated commitment to volunteering and social activism (the "Gray Panthers"), turn inward in a search for religious and spiritual meaning, or feel compelled to write an autobiography or memoir. An old-age style might also be signaled by a quest for personal growth through therapy, adult education, and travel. The elderly who seek unusual endeavors and set risky goals in novel and unexpected ways have radically reconceptualized and restructured their thinking, compared to their youth and middle age. Laypersons who have a remarkable late-life change are "Ulyssean," named after the Greek hero of mythology whose adventures occurred late in life (McLeish, 1976). Aging adults with an old-age style may also appear cranky, irascible, grumpy, and unpredictable. But like artists, they are misunderstood, their behavior is unfairly labeled pathological, and they are erroneously diagnosed as having a mental breakdown, being senile, or "not acting their age."

To better understand aged mavericks, whether they are researchers, scholars, or laypersons, the same questions asked of aging artists are applicable: Are the changes good or useful (quality)? What are the results (quantity or productivity)? Who are they (identification)? What are their characteristics (description)? In how many is this new behavior found and how often does it happen (quantification)?

An old-age style might be profitably looked for among amateur scientists, scholars, and artists. The status of amateurs falls somewhere between the professional and the layperson, but the latter are probably easier to recruit for research than the former, since they are not as busy. Signs of the old-age style, though, may be more muted among amateurs than scientists and scholars but more evident than in laypersons. "Sunday painters" are probably better candidates than scientists and scholars because they are more numerous.

A Caution.   Just as not all late art reflects creativity or an old-age style, not all eccentric behavior in the aged is valuable. A certain amount of wariness is therefore warranted when evaluating unexpected changes in older people's thought, work, and action. Growing old for anyone, artists professionals, and others, has varied consequences, of which only some are positive. The old-age style in the general population, as it is among historical artists and probably for contemporary artists, is as infrequent as wisdom and as rare as late-life creativity, and as difficult to discern as both. Marked shifts in the actions of older individuals are likely to be labeled "crazy" rather than good or creative. But a greater recognition of the old-age style might lead to a reconsideration of older people who have unfortunately and mistakenly been labeled quirky elders.

Investigations of marked late-life changes in the arts, professions, and the general population need to fulfill the same basic criteria as the study of painters did: the individuals have to be long-lived and still active. Living long provides a larger number of opportunities, more time to work, a chance to improve on past failures, and gives others a chance to discover, encourage, and applaud late-life developments. A long life also allows for knowledge, skills, and experience to be accumulated, which were major factors in the creativity of contemporary aging artists.

However, most people retire and stop being active in their occupations when fairly young. Retirement occurs for many by age 65, and at increasingly younger ages, sometimes involuntarily ("downsizing"), and late-life career shifts to new fields are difficult and infrequent. Old scholars leave the stacks of libraries, aging scientists close the doors of their laboratories, and senior citizens move to gated communities and play golf. For these retirees, there are few or no occasions for an old-age style to emerge. For them, an old-age style as a revitalization of late life is not likely to appear, at least as traditionally conceived.

INTERDISCIPLINARITY

The old-age style ties together art, art history, and creativity with psychology and gerontology. Cooperation between these disciplines occurred, fruitful exchanges took place, conflicts were surmounted, disparities in methods were reconciled, and goals converged (Chapter 2). Exchanges between artists, scholars, and scientists were possible because the topics, late-life creativity and the old-age style, were of mutual interest. Empirical investigations presented scholars with facts, narratives of individual artists and specific works were placed within a scientific context, and reports by aging artists on how they met, overcame, and transformed the effects of old age illuminated the study of creativity, psychological development, and aging.

Concrete examples of exchanges across disciplines were numerous. Art historians selected artists and descriptors of the old-age style (Chapters 12 and 13). Importantly, their choices were neither so idiosyncratic as to make the surveys unusable nor so stubbornly intractable and wildly divergent as to be unmeasureable. Despite the difficulty of the tasks, humanists fairly quickly nominated several hundred artists and dozens of terms on a controversial topic. The experts were not immobilized by all they knew about the old-age style or the opinions of other authorities. Signs of indecisiveness, the "undecided" choices, were infrequent. Consensus was possible because the historians were given specific instructions on what to do; how to make selections was clearly explained. Unlike wide-ranging discussions in the literature, often fuzzy, obscure, and esoteric, lists of names and descriptors were presented in rather prosaic but workable checklists, the responses were tallied, and compared to the published literature. No complaints were voiced about oversimplifying complex phenomena, of opinions being inadequately captured by a simple yes–no vote, or of personal views being constrained by scientific procedures. Professors of art performed as well as undergraduates under restricted conditions.

Similarly, artists gave useable reports about their late-life creativity despite the restraints of a questionnaire. Contrary to reputations for rampant individualism, artists dutifully completed a lengthy open-ended questionnaire on the impact of physical-sensory losses and ill health on the art of their old age (Chapters 12 and 13), again without complaint, as they did for several aspects of creativity, including numerical rating of the quality and quantity of their past, present, and future work (Chapters 9 and 10). Uniquely private opinions of artists were filtered through the formal procedures of content analysis, thereby transforming highly individual comments into statistical summaries. Like the responses of art historians to checklists and lay judges to the unique qualities of pairs of paintings, the artists' reactions were not so ambiguous or arcane as to make analyses impossible. The contributions of artists to empirical inquiry, like those of scholars, were neither ignored nor trivialized (Chapter 2).

The reining in of the specialized sensibilities of art historians and artists, and the harnessing of idiosyncratic discursive material, attests to the possibilities of interdisciplinarity across the arts, sciences, and humanities. Art historians and artists were able participants in scientific investigations that made several major contributions to the empirical literature. The scientific method, despite its unnatural demands and artificial settings, was successfully integrated into humanistic and artistic frameworks. Scientific inquiries could be played out against the backdrop of art and artists, and the outcome was complementary rather than antagonistic. The myth of the intractability of art to scientific analyses was fractured, as was the incapacity of scholars and scientists to share their expertise. An empirical perspective meshed depthful narratives of the single case and uniquely personal artistic intuitions. Scientific, humanistic, and artistic approaches, working together, achieved a greater understanding of the old-age style than one discipline by itself could. These exchanges, moreover, refuted a pathological-disease model of old age and a view of creativity that emphasizes declines and losses.

The juxtaposition of group studies, favored by science, with individual analyses, the forte of the humanities, along with the expressive strengths of the arts and artists, illustrate the possibilities of interdisciplinarity. Broad quantitative generalizations based on group data were qualified by an individual analysis of artists and works. Thus, the general effectiveness of five types of judgments for pairs of art was broken down for different kinds of art (figurative, non-figurative, and abstract) and for specific artists (Chapter 12). What holds in general was a context against which particular kinds of art and specific artists were placed. The skepticism of humanists to scientific inquiry was therefore mollified.

Scholars and artists are not only useful sources for research materials, study participants, and ideas for future research. As "workers in the field" with "hands-on experience," they are also in a good position to evaluate the outcome of empirical efforts and to suggest improvements. As scientific outsiders, scholars and artists can advise researchers if the restrictive procedures and unusual demands of research that disentangle extremely complicated events from one another misuse art or distort their original intent.

Empirical methods and statistical skills, for their part, clarified murky scholarly observations and obtuse artistic imaginations. The jargon of art historians and the exotic insights of artists were translated into testable facts. Historians and artists can return the favor by indicating what science has left out. In this way, there is a cycle of give-and-take between disciplines that complement, modify, and supplement quantitative, narrative, and expressive modes of inquiry. These exchanges counter truncated perspectives on psychological development, aging, and late-life creativity. The inclusion of artistic, scholarly, and scientific sensibilities into the study of the old-age style should increase interdisciplinary awareness and acceptance of each other's goals, especially when they are directed at a common problem, like aging.

This hope is nurtured by unexpected findings that flowed from these interdisciplinary efforts. An example is the preference for large canvases preferred by older artists. It is a choice that is congruent with post-formal thinking, a concept discussed in cognitive and developmental psychology. Post-formal thinking refers to a major shift in ideation with increasing age. In youth, formal thinking is ascendant, and the emphasis is on accuracy, specifics, and details. But with age, mental operations become less differentiated and more global, which are ways of describing works by aging artists with an old-age style. (For a discussion of post-formal thinking in the context of creativity, see Sinnott, 1998; in the context of aging, see Clayton & Overton, 1970; and for a more general discussion, see Commons, Richards, & Armon, 1984.)

The broad kind of thinking that characterizes post-formal thought converges with another general phenomenon of aging, the elderly's propensity for extracting the "gist" of an event rather than its details, on which younger individuals excel. For older artists, and in particular those with an old-age style, gist transforms complicated ideas into their essential or unitary themes, and these are expressed in late works. In contrast, youthful artistic works (before the advent of modern art, at least) should be more literal, showing off the artist's skill, and "true to life" in its details. Where younger artists might therefore paint historical works, landscapes, and still-lifes, older artists with an old-age style would produce works that are more abstract, symbolic, and expressionist, that is, post-formal in spirit, and capturing the gist of the artist's message. The same filtering process may be at work in modern art, which represents a later stage of development in art.

CONCLUSIONS

An understanding of the old-age style of artists and art contributes to the study of creativity and aging in general, draws attention to changes in late-life development that are unusual, unexpected, and positive, discloses the surprising strengths of aging, and brings disparate disciplines closer together. The old-age style is also relevant to everyday life. Producing a different kind of art in old age requires a number of complex abilities and traits: inspiration, conceptualization, concentration, and application; a vivid imagery and a rich imagination; and problem-solving skills, to name a few. These qualities are germane to any endeavor undertaken late in life, not just art. The old-age style therefore illustrates the oft-heard but seldom demonstrated virtues of growing old. The phenomenon also offsets the typical overemphasis on declines with age, thereby encouraging a reevaluation of the overlooked possibilities of aging. Loss is not the only consequence and stability is not the best to hope for. Talents continue, skills are reinvigorated, new ideas emerge, and life takes new directions.

But marked shifts late in life are rare, even for long-lived and active seniors, and characterize only a small fraction of them. But even if the old-age style were limited to a handful of artists, the interests of the aged are served by learning more about it. "Careers of older artists who continued to work well into their later years [are] a testament to older Americans' cause," asserted Byers (1982, p. 1). For Butler, genius is ageless and "creative imagination is not limited by time" (quoted in Byers, 1982, p. 2). Picasso is reported to have said (quoted in Greenberg, 1987, p. 28), "Age only matters when one is aging" but then continued, "now that I have achieved a great age, I might just as well be 20." The old-age style demonstrates that successful aging is possible. (For the tie between successful aging and wisdom, although not specifically with creativity and the arts, see Baltes, Smith, Staudinger, & Sowarka, 1990.)

Knowledge about aging artists, old art, late-life artistic creativity, and the old-age style is therefore not just about art history, extraordinary artists, and great art. The possibility of positive changes with age will make it difficult for agencies that serve an aging public to conclude that mental and other abilities inevitably decline in the later years. Further, the potential benefits of aging encourages a personal view of growing old that is optimistic. While aging artists with an old-age style are only a handful, they are like any elderly person burdened by the handicaps of aging, the aches and pains of arthritis, the losses in sight and hearing, the debilities of illness and poor health, the slowing down of mental abilities, and frailty. But artists with an old-age style demonstrate that compensations can overcome these debilities. Consequently, these artists are a model of what might be accomplished by the elderly, albeit in more modest contexts. In addition, aging artists with an old-age style prove that higher order cognitive and other abilities, talents and skills, the motivation to excel, and the ability to do productive and valued work can be maintained, strengthened, and revised, if not established for the first time in old age. If old artists can accomplish so much despite aging's pitfalls, others can too. Thus, the study of aging artists with an old-age style, as well as investigations of artists whose creativity was sustained late in life (Part III), affirm the vitality of aging.

The old-age style is an exemplary form of successful aging in the creative realm (Chapter 1), and is the artistic counterpart to wisdom. Common to both the old-age style and wisdom, as well as late-life creativity and successful aging, is their rarity, positive connotations, and late emergence, as well as a higher incidence of speculation than proof. In short, studies of the old-age style among artists bring to the fore several important yet relatively unrecognized facets of late-life achievement.

Art therefore has the revelatory power to uncover the strengths of old age, including late-life creativity (Part III) and the old-age style (Part IV). This ability of art can also be demonstrated in more ordinary ways, among less extraordinary elders who are not artists. That possibility is the task of the next two chapters of Part V. In Chapters 15 and 16, the overriding question was, as in previous chapters, How well do the elderly perform when faced with art?

PART V

Art and the Elderly

I'm growing fonder of my staff;
   I'm growing dimmer in the eyes;
I'm growing fainter in my laugh;
   I'm growing deeper in my sighs;
I'm growing careless of my dress;
   I'm growing frugal of my gold;
I'm growing wise; I'm growing—yes—
   I'm growing old.
              J. G. Saxe, *I'm Growing Old*

CHAPTER 15

Reactions to Paintings by Older and Younger Viewers

It is always in season
for the old to learn.
Aeschylus, *Fragments*

Studies of contemporary and historical old artists and late-life paintings (Parts III and IV) throw new light on the course of late-life creativity, bring attention to a special feature of artistic creativity known as the old-age style, and offset pessimistic perspectives on aging, among other consequences, including the furthering of interdisciplinarity (Chapter 14). The relationship between aging and art is examined in this and the next chapter again, but this time more broadly, for its relevance to non-artists. Their reactions to paintings, examined in this chapter, and more generally, their participation in the arts, along with their attitudes and interests toward a variety of art activities, in Chapter 16, bear on both late-life abilities and late-life creativity.

CREATIVE EXPRESSION AMONG AGING NON-ARTISTS

Looking at a painting in a museum is similar in certain ways to how artists create a work of art. Viewers put themselves in front of the artist's easel, so to speak, and struggle to make sense of the creator's vision and its particular expression. What was the painter attempting to achieve? Similarly, what is the sculptor, actor, or dancer communicating? What is the writer saying? Non-artists' empathy with the creative process and intuition about the artistic product, whether in response to a painter's canvas, sculptor's statue, actor's gestures, or a dancer's choreography, evoke unusual ideas and unique feelings that are rarely matched by other objects or events. Viewing art therefore touches on something akin to and roughly parallel with the creative act. Elderly art audiences, like aging artists' late-life creativity, wrestle with problems to be discovered, defined, and solved (John-Steiner, 1995; Richards, 1999; Runco & Dow, 1999).

Old non-artists' reactions to art should therefore open a window to late-life abilities, just as art reflected the strengths of aging artists (Parts III and IV), about which, as for aging non-artists, there are doubts (Chapter 1). Elderly viewers' reactions to paintings indicate adaptations and adjustments to physical, sensory, cognitive, and other losses, just as the accomplishments of old working artists did.

The elderly's best, though, cannot come to the fore in the typical research settings in which they are tested. Innocuous laboratory materials, arcane and irrelevant, are neither interesting nor appealing. A fast reaction (often in milliseconds) also hinders optimum performance, as does having to sit still, often in awkward positions, made more difficult by aging's ailments, like arthritis. Elderly research participants are less familiar and comfortable with artificial and controlled investigations than undergraduates, who have recently volunteered to participate in similar research, heard about studies in classes, and read about them in textbooks. The young are also less likely to mind being told what to do by an authoritative (and authoritarian) laboratory assistant who may also be their classroom teacher and not much older than they; they have more or less docilely followed orders in school and at home for years, and it is easier to subordinate themselves to older figures. The young are therefore more at ease than the elderly under the restricting conditions of the lab, more willing to conform to and accept the rules, and more obedient to arbitrary demands (which is not to say that they like them). Negative feelings, if they occur, are offset by the opportunity to earn class credit for participating in a study, often required to pass a course, and by experiencing first hand the work they have heard or read about. For undergraduates thinking of starting a professional career in science, participating in research furthers their ambitions.

Age-related differences in certain aspects of intelligence also put the old at a disadvantage. "Fluid intelligence," at which the young excel, refers to the ease with which information is accessed and manipulated, the quickness with which strategic shifts in learning are carried out, and the ability to select efficient techniques to facilitate remembering (Horn & Cattell, 1996). The elderly, in contrast, have the advantage of "crystallized intelligence," the knowledge that comes with increasing age and experience. But in the laboratory there is no time, opportunity, or encouragement to draw upon this storehouse of accumulated knowledge.

Tense and anxious, intent on doing well yet laboring under the uncomfortable circumstances of being a "subject," the aged do not accommodate as well as the young to laboratory pressures. Consequently, older participants feel inhibited and unenthusiastic if not indifferent, annoyed, frustrated, angry, worried about embarrassing themselves, and preoccupied over doing the right thing, not saying the wrong thing, and responding correctly. Uncomfortable and stressed, older research participants are not given the time to reflect or turn to alternative modes of thinking honed over a lifetime of experience. They are neither instructed nor encouraged to be expansive or deliberate, and even if they were, simple materials do not lend themselves to openness and flexibility. It should therefore come as no surprise to find the elderly doing more poorly than the young in research.

Even when a study is completed, the situation does not improve for the elderly. Questions and comments are hurriedly covered by the researcher in the expectation of the imminent arrival of the next scheduled subject. Undergraduate participants are grateful, since they can leave and reach their next class or appointment on time. Even if more time were available, participants are rarely fully informed about the task in which they participated. Too much knowledge, if passed on to future participants, would "contaminate" the results.

Procedural barriers to the elderly's optimum performance are neutralized with art. Art is attractive. It captures attention, piques interest, provokes images, evokes meanings, challenges accepted standards, and stirs feelings. Multiple reactions to the ambiguities and unclarities of art cannot be confined to a fast, single, or "correct" response. Yet these responses to art, like those to ordinary laboratory materials, indicate what is perceived, thought about, and felt. But with art, the stimuli are not artificial or simple, and the response is similarly complex and rich. The "crystallized intelligence" of the old can therefore be put to good use with complicated materials like art that also encourage free and open-ended communication. But the provocative qualities of art are the very reasons it is avoided in scientific research. Art is a "messy" stimulus whose response cannot be easily controlled or measured. Hence, nonsense shapes rather than abstract art, sandpaper swatches instead of marble surfaces, isolated auditory pitches replace musical themes.

The elderly, though, can take advantage of the distractions of art. The multiplicity of reactions—mental and emotional, sensory and perceptual, motivational and social, personal and attitudinal—are options from which the old can select the ones that make most sense. When "stuck"—what does this stimulus mean and what can I say about it?— alternatives are available. Older research participants welcome this profusion of possibilities, benefiting from having more than one way of maneuvering around age-related limitations and replacing them with other competencies.

Because art is less restrictive than normal laboratory materials, it is a fairer test of cognitive and other abilities of older subjects. Consequently, aging non-artists can perform at their best, and fully utilize imagination, flexibility, imagery, inventiveness, and problem-solving experience—the same kinds of abilities and strategies taken by aging creative painters in overcoming artistic dilemmas. The abilities of older people are therefore better revealed with art than traditional laboratory stimuli, and as a consequence, they might do as well as younger participants or at least not as poorly as they do with impoverished materials.

Art is seldom used, though, in investigations of aging. Most studies that do use art recruit adolescents, children, and college students, rarely the middle-aged, and hardly ever the old (Gardner, 1973, 1974; Taunton, 1982; Winner, 1982). The research concentrates mainly on the kinds of art preferred or the place of art in the school curriculum, rather than on general issues of aging (Ahmad, 1985). An exception is Parsons' (1987) illuminating discussion of the evolving responses to art with increasing age. Over 300 children and professors were interviewed on their reactions to paintings by Picasso, Goya, Renoir, Bellows, Klee, Chagall, and Albright. Through adolescence, Parsons found, viewers emphasized subject matter, beauty, realism, and emotional expression. But with increasing age, greater attention was given to the medium, form, and style. For older viewers, at the highest stage of art appreciation, autonomous, critical, and personal judgments predominated. Parsons was not explicit, though, about the exact ages at which these stages occurred, how they were achieved, or how many older people reached the highest level.

The paucity of art in gerontological research makes it hard to predict how poorly or well older participants will do compared to younger viewers. Age-related differences should be found since the old and young differ in so many ways. Thus, preferences for different kinds of art might be related to modes of thinking characteristic of each age. The elderly might prefer pastoral scenes, still-lifes, and landscapes for their contemplative inspiration. The young, with more active and energetic temperaments, are likely to be intrigued by unruly city scapes, frenzied historical events (rape, pillage, war), turbulent sea-scapes, and romanticized and mythical portrayals of love and courtship. There might also be age-differentiated reactions to portraits of younger and older men, and depictions of lusty women and nudes. It is also possible that scenes with older people might evoke younger viewers' prejudices, ageism, and stereotypes while older observers would react more benignly. Divergent outcomes could also arise from the elderly's more conventional attitudes and teenagers' rebellious spirit. Thus, older viewers would react more favorably than younger ones to traditional (representational) and religious art, while the latter would probably favor abstract, avant-garde, and surrealist works.

The broader question is this: Is there is an age difference in the response to art? More pointedly, do the young outperform the old? There are good reasons to predict they would. Younger viewers have better visual acuity and color discrimination than the old, making it easier for them to differentiate the intricate forms, subtle colors, and other details of a work. The young also respond faster and more accurately than the old.

Older viewers have advantages, too, and these may balance out youthful strengths. Take the reduced ability of the old to see details, a liability, but it is compensated by their proclivity to extract the "gist" or essence (Commons, Richards, & Armon, 1984). Older viewers can therefore achieve a more global understanding of a work of art, arrive at a general impression of its broader meaning, and gain a firm grasp of its overall mood. Younger observers miss these larger qualities because they dwell on a work's particulars: the specific objects, colors, and shapes, the play of light and shadows, and the lines of space. By over-concentrating on marginal effects, secondary figures, and peripheral objects, the younger viewer's eye is distracted from a work's larger theme. In other words, youthful viewers see the proverbial trees and not the forest. But older viewers can establish the essentials or gist of a painting, especially if they have the time to probe its multiple levels of meaning. Hence, they inadvertently follow good strategies for understanding a work of art.

Elderly viewers may therefore do better than younger ones in absorbing and unraveling the complexities of art. With art, taking more time to look and think gives elderly observers a greater and deeper understanding than youngsters' hurried glances. Older viewers, by taking the time to reflect on a work of art, have more images, ideas, and feelings than the more efficient but fleeting looks of youthful audiences. Slow old-timers may therefore gain a more profound and longer lasting reaction to art.

There are therefore many reasons to believe that young and old viewers will react differently to art, and an argument can be made for either one or the other age doing better. A third possibility: age differences will not be found. The old are favored over the young on some aspects of certain kinds of art, responding more fully and deeply, but doing poorly when speed, accuracy, and efficiency count. Put another way, the advantages and disadvantages of each age group cancel out as age-dependent handicaps are balanced by age-related strengths. But even if younger and older viewers' responses to art are equivalent, aging "wins." Parity indicates that the sensory, cognitive, and other losses of aging are not so severe as to degrade reactions to materials as complex as art. These possibilities were examined in the studies that follow.

THREE STUDIES

Age differences were examined in three unpublished studies on a fairly wide range of reactions to several kinds of art by different artists. The studies addressed the following questions: (1) Do younger and older viewers respond differently to art? (2) If so, are young viewers "better" (e.g., faster, more complete, greater accuracy) than old participants—or is the reverse true? Or are both the same? (3) Whatever age group is favored, does this superiority (a) hold for all kinds of art, representative as well as abstract, and for all artists, a van Gogh as well as a Mondrian; and (b) do differences in a painting's reported details, complexity, amount of skill, preferences, and the like differentiate between young and old viewers' sensory, perceptual, cognitive, affective, and other skills? Finally, (4) Are differences in responses to art less a matter of age than of gender and background in art?

Senior citizens and college undergraduates' reactions to art were compared on the above five questions. The latter was composed of 21 students, ages 17–22, who were volunteers from several undergraduate introductory psychology classes at a university in a medium-sized city in northwest Ohio; they were participating for course credit. The older (60+) group was composed of 18 attendees at two senior centers in the same city from which the undergraduates were recruited. The participants individually examined a series of paintings, and took as much time as needed. The paintings were postcard-sized reproductions, a format that allowed them to be easily handled in noting details and other features (in Study 1) or to sort them by styles (in Studies 2 and 3).[1] The three studies immediately followed one another. Prior to the first study, members of each age group completed a six-item

---

[1] On the general method: The participants' responses to the questionnaire were converted to a 4-point scale and summed over six questions, yielding an overall art background score.

questionnaire on their background in art. The scores for the two age groups did not differ, on average, and did not influence their responses in the three studies. Hence, one of the questions posed at the outset (#5) is answered: Experience with art, minimal in this sample, did not distinguish the responses to art by the two age groups (see footnote 3 on differences unrelated to age).

Age Differences in Reactions to Paintings

Art can be looked at in many ways (Lindauer & Long, 1986). For example, Russell and George (1990) had subjects aged 10–80 respond to 15 paintings on several "esthetic" scales, including likeability, pleasingness, preference, content, and style. The present study was more ambitious, and used 15 different scales, a number limited only by time constraints,. Time restrictions also prevented the participants from being questioned about their reactions, although the older respondents nonetheless spontaneously did so without being prompted. The materials and procedures were pretested in a pilot study.

The participants were shown representational (realistic) and abstract paintings by eight artists, with one of two examples for each, both of which were similar in subject matter and time of production. Each participant therefore saw eight paintings (of 16 available). For example, half of the viewers saw Rembrandt's "Self-Portrait" (c. 1636) while the other half were shown his "Portrait of the Artist's Son, Titus" (c. 1645–50). As it turned out, the responses to the two works by the same artist did not differ, and both sets of results were combined. The representational works, equally divided into landscape and figurative art, were by van Gogh, Gauguin, Rembrandt, and two anonymous mass-produced works. The abstract paintings were by Mondrian, Stella, Miro, and Still.

The art was rated in 15 ways. Two were related to acuity: (1) the amount of detail seen in the work, and (2) the ability to pick out the separate parts of the painting. Seven scales rated perceptual–cognitive abilities: (3) complexity, (4) familiarity, (5) novelty (or unusualness), (6) interestingness, (7) skill, (8) subtlety, and (9) arousal of ideas,

Continued
The representational paintings were shown first, followed by abstract art examples, a sequence that minimized an unwanted "startle effect" in reaction to non-traditional art's unfamiliarity, unusualness, and to some, bizarreness. The reproductions were shuffled before each presentation to insure that each viewer saw a different sequence of paintings. The task was explained and practiced with a single work.

thoughts, and memories. Three additional scales were affective: (10) liking, (11) the feelings or mood evoked, and (12) shock value. Two general judgments were also obtained: (13) the work's overall quality and (14) the certainty with which the judgments were made. The final measure (15) recorded the time needed to rate a work. The 15 measures always followed the same order. The participants described their reactions to each work with one of four words. When asked how skillful a work looked, for example, the choices were "very, moderately, slightly, or not [skillful]." In the case of the viewers' certainty, they could choose "strong, moderate, slight, or no [certainty]." The choices were translated into numerical values ranging from one to four.

How did the young and old do? Their reactions to the paintings were identical on all but one of the 15 measures; the exception was time to respond. Younger subjects, as is usually the case, completed the task almost a minute faster than the older participants (Mean=105.57 and 163.91 s, respectively). Except in this one instance, no other difference characterized younger and older viewers' responses to representational and abstract art by eight different artists.

Thus, sensory, perceptual–cognitive, and affective losses of aging were not apparent in viewers' responses to art. Details were reported to be equally seen by both the young and old; the ideas, thoughts, and memories evoked by the paintings were similarly rated by the two age groups, as were feelings and moods; both young and old were equally certain of their opinions. In short, losses were not apparent in the older group's ability to differentiate, discriminate, and evaluate art.

When the type of art was taken into account, though, a few age differences were found on four of the 14 measures (29%; time-to-respond was excluded since it was already shown to differ with age): Older viewers liked representational art more than younger viewers (Means = 1.97 and 2.36, respectively), while the opposite was true for abstract art (Means = 2.83 and 2.60, respectively). Elderly respondents also found representational art more subtle and interesting than abstract art. For younger viewers, either the reverse was true, or judgments were the same for the two types of art.[2]

Except for these four age differences, which depended on the type of art, and the time it took to complete the task, young and old viewers reacted identically to the paintings; they were much more alike than

---

[2] Statistically significant differences were found for time ($F(1,30) = 24.63$, $p < .01$) and between the two types of art on four scales ($F(1,30) < = 5.64$, $p < .03$): interestingness ($F(3,90) = 28.55$, $p<.01$), liking ($F(3,90) = 7.14$, $p<.02$), complexity ($F(3,90) = 5.64$, $p < .05$), and subtlety ($F(3,90) = 7.95$, $p<.01$).

different. Even younger viewers' faster response was not necessarily due to the older participants' attentional or other deficits. The difference reflected the different ways in which the two age groups took the task. As noted earlier in introducing the three studies, older viewers were more willing and eager to talk about what they were doing and to explain their reactions. Younger respondents, on the other hand, participating for class credit, completed the task quickly, presumably in a rush to get to their next class, the library, dorm, or coffee shop. In contrast, older viewers were delighted by the attention received from a university professor, were glad to be taken seriously, and were happy to give their opinions—even though not asked for. They may have also felt it was socially necessary to account for their reactions, either out of politeness or perhaps because of wanting to please the researcher. They were either more willing to share their reflections, or as retirees with no pressing obligations, they had the time to "schmooze." Whatever the reason(s), time-to-respond, the only general age difference, did not necessarily denote sensory or cognitive losses among the older participants.

The few differences between young and old viewers as a function of the type of art—on preference, subtlety, interest, and complexity for representational and abstract works—could also be interpreted in several ways and not necessarily as signs of the superiority of one age group over the other. On the one hand, the older viewers might have preferred representational over abstract art because sensory deficits associated with age made it hard for them to deconstruct the amorphous forms and ambiguous details of the latter, unlike the former type of art's clearer and familiar depictions. On the other hand, the difference could simply reflect the taste and experience of each group with one or the other type of art. Older viewers may have found representational art more interesting and understandable than abstract works because they were more familiar; it predominated in their youth when modern art was less popular and accepted. Similarly, younger viewers' reactions to abstract art might reflect a greater ability to comprehend complex material, or alternatively, a comfortable feeling with unusual, bizarre, and strange depictions that reminded them of psychedelic CD jackets and rock stars posters. Age-related differences in preferences for different kinds of art were also reported by Ahmad (1985). Avant-garde (modern, abstract) art was less popular among children than traditional art, and subject matter and content had less influence on preference as age increased (through college) than "more complex stimulus dimensions such as style, color, line quality, texture, and other aspects of composition" (p. 103). Whether older viewers are more like children or young adult in this regard remains a subject for further study with a broader range of ages.

It is also possible that equivalent reports did not indicate what young and old viewers actually saw, understood, or felt. Consider the reported absence of an age difference in perceiving the details of a work. The older group might in fact have seen less detail, contrary to their self-reports, had they been objectively tested for accuracy. Self-reports, despite their closeness to experience, have their limitations (Part IV).

There are obviously more than 15 ways of reacting to art, and the ones that might tap age differences were not included in the study (Lindauer & Long, 1986). For example, had more technical questions been asked, say about forms, light, and color, younger viewers might have done better than older viewers because of their keener eyesight and better color discrimination. Unexamined, too, was style, a major attribute of art (Part IV). The next two studies investigated differences between young and old viewers in the perception of paintings' styles.

Age Differences in Distinguishing the Styles of Representational and Abstract Artists

Style is not an obvious, specific, concrete, or denotable aspect of paintings, like the amount of detail or subject matter, which is why Hess and Wallsten (1987) call it an "ill-defined category." (For style recognition in modern poetry, see Arcamore & Lindauer, 1975.) An artist's style is discerned from subtle features of the brushstroke, colors, forms, and other qualities in many works that make them distinct and unique. Once these noteworthy features of an artist's style are identified, they then are remembered and compared with the distinguishing styles extracted from other artists' works. The detecting and abstracting of similarities and dissimilarities within and between works by different artists, retaining this knowledge, and transferring it to new works are challenging tasks that are neither completely conscious nor easily verbalized. Identifying artists' styles is therefore a extraordinary cognitive accomplishment. Does it depend on age?

Few studies have examined this question in older populations, at least with objects as complex as art. A notable exception is Hess and Wallsten (1987) who studied age differences in the use of 10 categories related to style (e.g., level of detail, color) for 20 paintings by the impressionists Monet and Pissarro. The authors found that younger participants relied on abstract categories and older viewers depended on concrete dimensions, such as similarity of content. Sharper age differences emerged in a second study, where the age groups were compared on the time needed to learn the names of the paintings' artists and apply them to a new set. The younger viewers did better, which the authors again attributed to their ability to abstract information, that is, to select the essentials from an array of complex information. The older subjects took longer, because they depended on the concrete. Thus, the sorting of the styles of unidentified works by several artists should take longer for older than younger viewers and be less accurate.

Accordingly, young and old viewers (the same ones who had served in the previous study) were shown sets of postcard-sized reproductions by six artists (not seen earlier). Three were representational—Cézanne, Manet, and Prendergast—and three were abstract—Diebrenkorn, De Kooning, and Kandinsky. For example, the three paintings by the American representational artist Prendergast (1859–1924) were "Handkerchief Point" (1897), "Umbrellas in the Rain, Venice" (1899), and "Bathing, Marblehead" (1897); and the three abstract works by the Russian artist Kandinsky (1866–1944) were "Open Green" (1923), "Sign" (1925), and "No. 398: Heavy Circles" (1927). The representational examples for each artist were similar to one another in subject matter, and the sets in both types of art were painted within a relatively short time of one another. These similarities minimized extraneous clues, other than style, that might distinguish the works.

Young and old participants were first shown a set of nine examples of representational art by the three artists, three examples of each, followed by nine abstract examples by three other artists, or a total of 18 works. They first sorted the nine representational works, followed by the nine abstract examples, into groups that appeared to be painted by the same artist. Participants were not told how many artists or examples of their work were shown. They were free to manipulate the postcard-sized art in any way they chose. Some laid them all down on the table and worked on them simultaneously, for example, while others looked at each work, one at a time, and then laid them out. Changes in the sorting could be made at any time. In cases where no similarities were seen or the works could not be sorted for any reason (none had to be given), the art was placed in a separate pile. As much time as needed was taken, a procedure that made this measure too variable to be recorded.

How did the young and old do on this task? The accuracy with which the works were correctly sorted by style did not differ between the two age groups, paralleling the first study where hardly any age differences described the responses to art. The sortings by the young and old were not only equal in accuracy but high (Mean=2.48 and 2.23, respectively; three is the maximum number that could be correctly sorted). In addition, both representational and abstract art were sorted at about the same—and high—level of accuracy. Further, the six artists did not differ from one another in the number of works successfully sorted, with one exception: Manet's works were more accurately sorted than the other two representational artists, but this was independent of age.[3]

Younger and older viewers were therefore nearly perfect in assigning the correct styles to 18 paintings by six representational and abstract artists. Categorizing and abstracting, two major cognitive skills upon which the formation of style depends, were not age-dependent. While Hess and Wallsten (1987) found age differences, their study differed from this one in several important ways: They used more works but by only two artists; they relied on judgments of similarity rather than sorting behavior; they focused on the reasons for the choices rather than accuracy; and finally, all the paintings were representational. In contrast, when abstract art was used in a study of recognition rates by Harker and Riege (1985), age differences were irrelevant, as was the case in this study. Age differences in the response to art awaits further investigation.

One direction is to make the task harder. With both age groups performing near the maximum, there was no opportunity for age distinctions, if any, to differentiate the two age groups. Had more artists and examples been shown, making the sorting of styles more difficult, an age difference might have emerged. The task could also be made more difficult by testing viewers' ability to distinguish style differences *within* the same artist's works, rather than, as in this study, *between* different artists' works. That is, viewers would look for differences between Eakins's earlier and later works, rather than contrasting the works of Eakins and Tiepolo. Accordingly, the next study examined differences in the perception of the styles of earlier and later works by the same artists.

---

[3] The combined young and old works by Manet were sorted more accurately than the other two representational artists ($F(2,54)=8.20$, $p<.01$). Art background affected the accuracy with which participants sorted the art, but independent of age ($F(1,27)=5.10$, $p<.03$): Viewers with more experience with art did better at sorting abstract art. Irrespective of age, too, abstract art was more accurately sorted than representational art (Mean=2.44 vs. 2.28, respectively; $F(1,27)=7.02$, $p<.02$). However, men and women participants, in general and irrespective of age, did not differ on the sorting task.

Age Differences in Identifying the Styles of Youthful and Older Art

Painters' styles often change over their lifetime, although not necessarily as radically as in the case of the old-age style (Part IV). Viewers' ability to detect these changes may depend on their age. Young viewers, like young artists, have a fresh and energetic perspective, focus on details, and prefer concreteness and clarity over ambiguity and fragmentedness (Chapter 13). Consequently, young viewers should be more alert and sympathetic to the content and treatment of youthful art, paralleling the sensitivities of young artists. Young viewers should therefore do better than older viewers in distinguishing younger from older works.

Older viewers have an advantage, too. Like old artists, the elderly have grown in knowledge and matured in thinking. They may prefer, like artists with an old-ages style, a rough brushstroke, a diffuse handling of light, and an amorphous treatment of forms. Thus, older viewers would be more sensitive to and empathic with the sketchy and incomplete qualities of late-life art. Both old artists and old viewers might be more open to and accepting of the uncertainties of the later years and the immanence of death, compared to their youthful counterparts, and late art may therefore be imbued with these qualities. Further, older viewers, like aging artists, resonate more fully to the gist of a painting, its overall meaning or "story," rather than its details and accuracy (Chapter 14), as younger viewers and artists do. Thus, to the extent old viewers have certain tendencies in common with aging artists, they should be more perceptive to the styles of older art. Accordingly, the elderly should do better than younger viewers in distinguishing older from youthful art.

With one age group more sensitive to the age-related qualities of art than the other, each will do better in distinguishing either early or late works: young audiences will do well on identifying younger works and older viewers will be more accurate with older art. Alternatively, the advantages of each age group will cancel out the other and no differences will be found.

To test these possibilities, two age groups (who also participated in the two studies reported earlier) identified style differences within younger and older paintings of six artists. Young and old participants were shown postcard-sized reproductions of representational art by long-lived artists painted in their youth and old age. The artists were Cassatt, Degas, Homer, Monet, Renoir, and Turner (none were shown in the previous two studies). The number of works for each artist varied, depending on the availability of early and late works as well as the similarity of their subject matter. For example, the three younger works by the American artist Cassatt (1844–1926) were "Five O'clock Tea" (c. 1880), "Woman Reading in a Garden" (1880), and "The Cup of Tea" (c. 1880); her older works were "Woman Arranging Her Veil" (1886), "The Letter" (1891), and "Woman Bathing" (1891). (Cassatt stopped painting well before her death because of blindness.) Only Cassatt's portraits of women were selected, rather than her well-known works with babies and children, in order to minimize age-related clues.

The selection criterion, unfortunately, resulted in relatively few younger and older paintings for some artists. The number of examples ranged from a low of six (for Cassatt, Renoir, and Degas; three for each age period) to a high of 14 (for Turner); two painters were represented by 12 works (Homer and Monet; Median = eight works). Regretfully, too, not all of the six sets of artists' works could be shown to every viewer because of time limitations (this was, after all, the third study faced by the participants, and all had to be fitted within the hour made available). The time restriction affected the seniors more than the young, who took the time to talk about their responses. Since the number of works and participants were therefore small and variable, a minimum number of statistical tests could be done. Consequently, the results should be considered suggestive, although their congruence with the previous two studies, as will be shown, enhanced their credibility.

The order in which the six artists were shown was randomized, as was the sequence of younger and older works within each presentation. The artists' names and the titles and dates of their works were masked, although each set was identified as painted by the same artist. Viewers were instructed to sort the works according to the ones that seemed to be painted in the artist's younger or older years; there was also a "can't say" category. The judges used whatever sorting strategy they wished, as in the previous study of style.

Did age matter in this case? No. The two age groups did not differ, as in the previous two studies. The number of artists' younger and older works correctly sorted by younger and older judges was about the same—and low. Overall, for both age groups combined, the percentage of paintings correctly categorized ranged between 39% and 45%, which is not statistically different from chance. For older viewers, only 40% of the works were correctly categorized as either young or old, with about the same number correctly sorted for the young and old sets (39% and 41%, respectively). Younger viewers also did poorly, sorting only 44% of the works correctly, and with little difference between younger and older art (44% and 45%, respectively). Age was not a factor, either, when incorrect assignments were examined: older art was as often placed in the "young" category as younger art was placed in the "old" group. The number of "unsure" responses were also not (statistically) distinguished by age differences (Mean=1.50 and 0.61, respectively).[4]

The results for the French artist Renoir (1841–1919), which were seen by the largest number of viewers (20) and are therefore fairly reliable, are illustrative. His six paintings, equally divided between youthful and older works, were as follows: the younger works were "Flowers in a Vase" (c. 1866), "Le Pont des Arts, Paris" (c. 1868), and "Pont Neuf, Paris" (1872); his older works were "The Seine at Chatou" (1880), "Dance at Ougival" (1883), and "The Vineyards at Cagnes" (1906). The younger and older works were correctly assigned by 43% and 48% of the older viewers, respectively, and by 42% of the younger viewers for both the young and old categories.

Each age group therefore did *not* do well with art that corresponded to their ages. Nor was it the case that age-related benefits were canceled out by age-related disadvantages, that is, the young doing well on youthful art and poorly on older art, with the reverse true for the older viewers. In short, old and young viewers responded similarly to both old and young art; the accuracy of their abstracting and related abilities, albeit low, was equivalent. There was no evidence for age-related differences in the perception of style; younger and older viewers were no better than one another (or chance) in distinguishing art that matched their ages. Young judges in previous studies (Part IV) successfully made age-of-production distinctions between 24 prototypical young–old pairs (albeit in the wrong direction; old works were judged to be painted in youth). But they only saw two works by each artist, and each was selected by experts for their protypicality as young and old examples. Experts would probably do better at differentiating sets of young and old art, but since they are also likely to be familiar with them, they would probably sort the art in terms of their knowledge rather than according to perceived age differences.

But for the untrained in art, perceiving style differences *within* a sample of an artist's younger and older work was a difficult task, in contrast to the previous study of differences *between* artists' styles, which was too easy. The more difficult task could be made easier, and the results more revealing of age differences, if the series of paintings

---

[4] Statistically, $F(1, 36) = 2.16$, $p > .05$) for incorrect assignments and $t(36) = 1.44$, $p > .05$) for "unsure" judgments.

for each artist had more sharply diverged in the ages at which they were done. Examples could be taken, for instance, from the artists' first and last 10 years of their working lives, when they were in their 20s and 70s or 80s. Had examples from more extreme ages been available, young and old viewers' ability to distinguish early and late styles might have been found.

Thus, works by Cassatt and Degas encompassed 11 and 18 years, respectively, spans probably not long enough to show a change in style, if any. Works by Renoir, Monet, and Homer were considerably longer, spanning 40, 34, and 35 years, respectively, but even these might not have been enough. For example, the oldest paintings by the French artist Monet (1840–1926), painted 18 years before his death and many years after his earliest works, excluded his most distinctive late-life pieces, the "garden" series.[5]

Reproductions by artists with an old-age style (Part IV) would have been ideal, but these were not available as postcard-sized art, or were too few in number or differed in subject-matter to be included. The failure to find an age difference in the perception of style could therefore be due to the absence of an age difference in viewers, or simply the result of methodological shortcomings: the insufficient number of examples, age spans for works were too short, and the art did not sharply enough portray changes over time.

Conclusion. Young and old viewers' inability to detect differences *between* the styles of younger and older works indicates the old were not worse, a result that is consistent with the previous study where style differences *within* works by the same artists were not found, and were generally, congruent with the first study on the multiple responses to art where age differences were also absent for the most part. Thus, in three different studies, age was largely irrelevant in reacting to art. The abilities of the old were similar to the young, at least in response to paintings. To the extent age-related sensory, cognitive, and other differences exist, and older individuals are at a disadvantage, these do not affect the ability to manage materials as complex as art.

---

[5] Monet's latest works were "Poppy Field in a Hollow Near Giverny" (1885), "Haystack at Sunset Near Giverny" (1891), "Rouen Cathedral at Dawn" (1894), "Ranch of the Seine Near Giverny I" (1896), "Water Lilies I" (1905), and "The Ducal Palace at Venice" (1908); and his earliest works were "The Ridge at Argenteuil" (1874), "Camille Monet and Child in a Garden" (1875), "Boulevard St. Denis, Argenteuil, In Winter" (1875), "Woman with a Parasol—Madame Monet and Her Son" (1875), "Meadow with Poplars" (c. 1875), and "La Japonaise" (1876).

However, it cannot be concluded, on the basis of an *absence* of differences, that age has no effect on or relationship to the response to art. The failure to find an age difference in three studies (with a few minor exceptions in the first study) may not have much to do with age, per se, as with the particular kinds of materials and procedures used (or excluded) and the sample of participants.

With these caveats in mind, each of the questions posed at the beginning of the chapter is answered with a resounding "no": there were no age differences in the response to art in general, and none as a function of the artists, types of art, or kinds of responses. Further, differences due to gender and art background were independent of age.

These results are quite unlike the findings for mundane and simple non-art laboratory materials where the old typically do worse than the young. Here, the old did as well as the young. A similar paradox characterizes the findings for late-life creativity (Chapter 1 and Part II): Old non-artists do poorly on tests of creativity and cross-sectional studies but well on longitudinal studies, and old artists do not display real-world creativity (Chapter 3) or do (Chapter 6 and Part III). The discrepancy between the two kinds of results could be resolved by adding art to the armory of materials used in laboratory studies, and by including artists as research participants in investigations that rely on ordinary materials (these additions are discussed in Chapter 17).

The relationship between art and age can be explored in other ways. Comparisons might be made between the number of young and old amateur painters, musicians and composers, writers of poetry, and the like. Do age groups differ in their participation in the arts? Tallies on more passive activities can also be made for 20- and 70-year-old readers, arts audiences, museum-goers, theater attendees, and so on. Do the young and old differ in their involvement in the arts? Both participatory and observational kinds of arts activities could be compared with serious amateurs outside of the arts: golfers, bowlers, travelers, science buffs, backyard astronomers, bird watchers. In addition to behavioral differences, interests and attitudes of the young and old toward the arts can also be compared. Do the young or old have a greater interest in and more positive attitude toward the arts? These questions about the relationship between age and art among non-artists are examined in the next chapter. The same questions were posed, as they have been in one form or another throughout this book: Are there differences between the young and old in relation to the arts? If so, is youth favored? Or is the profile for aging non-artists as positive as it was for old artists? Further, what bearing do scientific studies of art have on aging and late-life creativity?

CHAPTER 16

Age Differences and the Arts

I love being old, and can't wait to be seventy.
John Cowper Powys, Letter to Nicholas Ross, 1939
(at age 67), author of *The Art of Growing Old* (1944)

The elderly responded to paintings in the same ways as the young (Chapter 15). But would they also do as well across the arts? How many old people paint or play an instrument, let us say, compared to the number of youngsters who do? Are there age differences in attendance at the theater and concerts, or visits to art museums? More generally, do interests in and attitudes toward the arts differ for the young and old? Comparisons between the young and old on these broad fronts extend the book's focus from artists and paintings to the arts in general. Nevertheless, the focus remains on aging, cognitive and other kinds of late-life development, and creativity in old age. If aging non-artists' relationship to the arts in general is as positive as their reactions to paintings (Chapter 15), and parallels the affirmative findings for old artists (Parts III and IV), then another foundation is laid for an optimistic outlook on aging.

THE ADVANTAGES OF THE ARTS FOR THE ELDERLY

The reactions of non-artists to art, whether young or old, parallel artists' creativity in some ways (Chapter 15). Similar parallels exist between non-artists who participate in the arts and professional artists. Amateur musicians and audiences at music concerts, for example, experience the creative product directly and the creative process of the composer indirectly. Audiences at a symphony, or a dance concert or live theater, ask themselves, What is going on? What does it mean? When they do so, they are engaging in two key elements of the creative moment: problem-finding and problem-solving (Runco & Dow, 1999; see also Mayer, 1999). In addition, both audiences and artists share an aesthetic puzzle: they try to make sense of a work's meaning and their feelings about it in the absence of well-defined guidelines and clear precedents (Lindauer, 1982). Furthermore, the figuring out process, by artists as well as audiences, is not routine, but on-going and provisional: memory is searched, relevant material is selected, feelings are examined, and extraneous thoughts are discarded, and what is left is organized and reorganized. Like artists, too, nonartists engage in and appreciate the arts for intrinsic reasons, for their own sake. "[P]artici-pants invest heavily in time and energy and gain some degree of competence [they] can point to with pride" (Kelly, Steinkemp, & Kelly, 1986, p. 531). Expressive activities in the arts for both artists and non-artists, whether pursued for their own sake, or for enjoyment and self-fulfillment, take precedence over instrumental goals of practical-ity, usefulness, and monetary gain (Hiemstra, 1982).

These creative resonances are shared by both young and old arts audiences, but the latter gain more. Participation in the arts requires less energy and strength than vigorous forms of leisure and recreation, like hiking and ballplaying, and therefore need not decline with age (Setlow, 1984). Art-related activities, in addition, can be performed and enjoyed at home, alone, quietly, while sedentary, and at convenient times that suit the elderly. Perhaps this is why activities that are "seden-tary or home-bound," Cultler and Hendricks (1990) noted, "exhibit a relatively equal frequency across the life course," and solitary activities increase in "frequency of participation with age" (pp. 173, 176).

The arts offer many other benefits to the aged. They are opportu-nities "to act effectively, to develop social relationships, to maintain and increase cognitive skills, to exercise choice, to practice freedom, and to develop competencies in coping with change." They also "enrich … lives [and] facilitate the expression of ideas and feelings." Additional bonuses are "increased sensory stimulation[,] the develop-ment of new skills, improved self-esteem, [occasions] to relate to others in more meaningful ways[,] increased … awareness of self … heightened responsiveness to [the] environment … authentic expres-sions [of] experiences [,] the opportunity to discuss and develop ideas[,] to share … rich and varied experiences in life [, and] prepare [the elderly] to take risks" (Scotch, 1989, pp. ix–x). Arts activities in which the elderly can participate number at least 30, ranging from the traditional (drama) to the unusual (finger-painting and puppetry; McMurray, 1989; see also Cahill, 1981).

Opportunities for the elderly to engage in the arts, furthermore, are available, accessible, and inexpensive at neighborhood senior centers or at nearby museums and colleges. In group settings, arts activities encourage interactions with others, increase social contacts and the sharing of experiences, and are occasions for companionship and the forming of friendships. Art as a social event is particularly relevant for isolated seniors without a spouse, or with relatives and friends living far away. Whether carried out in groups or alone, art activities help the elderly discover more about themselves, unsuspected talents, vaguely sensed interests, and what they might do (Erikson, Erikson, & Kivnik, 1986; Manheimer, 1987–88). The elderly's participation in the arts, furthermore, enables them to practice a wide range of mental and physical skills: learning and remembering, reading and conversing, getting around and finding one's way.

The variety of behavioral, psychological, and social skills promoted by the arts improve senior's physical health, sensory competency, and mental well-being. For these reasons, various agencies promote and subsidize senior art programs, workshops, and lectures, make available free or reduced-price ticket to arts events, and provide transportation to art museums, theaters, and concerts (Moody, 1982b, 1987). Art programs for the elderly therefore ensure an active and vital old age (Johnson, Prieve, & others, 1976).

LEISURE STUDIES OF THE OLD

The many benefits of arts activities for seniors, listed above, are largely based on informal and anecdotal accounts. There are, though, a few empirical studies on leisure and recreation that document what the elderly do, and some include the arts (often used synonymously with "culture"), although many are outdated (Kleemeier, 1961). Nevertheless, these efforts provide some support to the arguments presented above, and are models of what might be done in the future in establishing the importance of the arts for the aged.

Among the varied leisure activities surveyed among 72- to 82-year-olds, arts-cultural domains ranked high, comparable to social activities and community service (Kelly et al., 1986). The authors reasonably concluded that the arts as leisure were "neither [a] peripheral [nor] residual [activity] nor life transforming elixir [but] part of life" (p. 537). A questionnaire of 500 elderly residents (average age was over 73) on recreational activities found that a substantial number attended plays or concerts (38%) and visited museums (21%; Udell, 1979). Among the 32% who attended discussion groups and the 7% who took classes, some were probably related to the arts. Further, the two most frequently cited reasons for engaging in leisure activities, for pleasure and social contacts, were goals to which the arts contribute. In a poll by Lou Harris (*The Myth and Reality of Aging*, 1978) of 4,254 respondents, 2,797 of whom were 65 or over (66%), older respondents reported that the best years of their lives occurred at 65 and older, mainly because they had more leisure time. A large number of leisure activities were listed (869), but only a few were clearly related to the arts ("going to the theater," "books, museums, reading"), although others might be related (taking classes, going to the library).

The leisure activities of "all ages" were compared to those "over-60" by de Grazia (1961, table 5.4, p. 121). Among the 21 activities listed, little or no differences distinguished the two age groups in singing or playing a musical instrument; both activities, combined, ranked fifteenth. Similarly undistinguished by age were reading magazines or books, activities that ranked fourth and fifth, respectively. There were few age differences, too, in attending a play, concert, or opera (combined rank=20). The only apparent age difference (no statistical tests were reported) in activities related to the arts was listening to records; it accounted for 14% of the all-ages group and 6% for those over 60 (the activity ranked 7th).[1] This type of study merits replication with contemporary forms of leisure. Deserving of follow-up, too, is de Grazia's findings on money spent on the arts (table 5.3, p. 119), which was not broken down by age.

In a questionnaire study with theoretical overtones, Petrakis and Hasnon (1981) determined that individuals over 55, distinguished by cognitive styles, also differed in leisure activities. Cognitive styles were defined by the respondents' ability to perceive embedded figures, using a test of field-dependence and -independence, which the authors interpreted as passive and active styles, respectively. Four types of activities were examined, but only the one labeled "crafts" has some bearing on the arts. Hopefully, this type of study with theoretical implications will be expanded to include more specific areas of the arts, along with comparisons between younger and older participants.

The most extensive examination of the relationship between aging, leisure, and arts–cultural activities is Terman's classic investigation of gifted children, now in their 70s and older (Holahan, Sears, & Cronbach, 1995; all quotes are from pp. 175, 176, 191, 217–218). For these intellectually superior elders,

> leisure and avocational activities [are] increasingly important in the later years [because they] invoke shared interests and the demonstration of

---

[1] To put de Grazia's (1961) findings into perspective, little difference was found between all-ages and those over 60 in watching TV (57% and 53%, respectively), an activity that ranked first.

competencies[, create] a sense of belonging and self-worth[,] challenge the older person [to] maintain established knowledge and competencies [and] compensate for some of the loss, with retirement, of positive aspects of interactions with others in the work setting.

The gifted elderly's range of leisure activities did not fit "the stereotype of the gifted as one-sided in abilities and interests." The 70- to 80-year-olds were engaged in about 17 activities, on average, compared to about 12 for the general population. These fell into five broad areas: intellectual, social, recreation (divided into "active," such as travel, and "passive," like reading newspapers and books), and cultural. The latter included listening to recorded music, going to concerts, playing a musical instrument, creative writing, poetry reading and writing, doing sculpture, and attendance at theater.

Of these five areas, intellectual pursuits ranked highest, not surprising for a group selected for their intellectual superiority as children. Cultural–arts activities in general ranked lowest, although not all did. Thus, reading was engaged in by 98% of the respondents, and other cultural–arts activities ranged between 36–87% of the respondents. Some 37% reported listening to music. Other activities outside of the arts/cultural sphere had even lower rates of participation, ranging from 2% to 10%.

The elderly's activities in the five areas were also related to their physical health and psychological well-being; the latter referred to life satisfaction, happiness, morale, and the absence of psychiatric symptoms as measured by self-reports, tests of autonomy, positive relations with others, having a purpose in life, and social interaction. The cultural–arts area was positively correlated with both physical health and psychological well-being: .21 and .14 for men, and .20 and .11 for women, respectively (Holahan et al., 1995, table 10.8, p. 230 and table 11.2, p. 246). The numbers, while seemingly low, were statistically meaningful. Further, these apparently low correlations were higher than the intellectual relationships: .13 and .04 for men, and .11 and .10 for women, for health and psychological well-being, respectively; and nearly the same as the correlates for social activities. The only instance in which correlations were higher was "ambitions in whatever engage[d] in": .33 and .36 for men, and .30 and .20 for women, in relation to physical health and psychological well being, respectively. The authors concluded that engagement in leisure activities, including those in the arts–cultural areas, had positive physical and mental consequences.

Individuals who still reported commitment to [leisure] goals at an advanced age also reported greater psychological well-being...particularly [those that] reflected an involvement with [cultural and other] activities.

The elderly's satisfaction with the cultural–arts area, furthermore, was high, accounting for 65% of the men and 85% of the women. These percentages were higher than satisfaction with religion, where the percentages were 44% and 65%, respectively, although both were lower than the elderly's satisfaction with their children, 89% and 93%, respectively.

For intellectually gifted individuals over 70, as well as the elderly in general, and indeed any age, leisure activities involving the arts are an important form of recreation. Nevertheless, the empirical studies that demonstrated these virtues are few, old, not particularly about the arts, and either narrowly define arts activities or vaguely refer to them. Once again, as was the case for elderly painters, old paintings, and the elderly's response to paintings, little is known about the relationship between the aged and arts activities. A recent review admitted that "very little information is available" on leisure activities among the elderly, and furthermore, that research in gerontology and psychology is "limited" even though leisure is "central to [the elderly's] mental health and morale" (Tinsley, Colbs, Teaff, & Kaufman, 1987). The lack of empirical data is especially distressing, the authors maintained, since "past research [has] demonstrated that older persons have a substantial amount of discretionary time, that they regard this time as important, and that the ways in which they spend it are different from the way younger persons do..." (the above quotes are from pp. 53–54). The authors took up the challenge with a questionnaire on the leisure activities of 1500 individuals who were 55–75 years old. Like previous studies, though, most of the activities were not related to the arts (playing cards, watching TV, and attending social and religious meetings and groups) and only a few were: ceramics, dancing, and reading, and a handful of non-traditional forms, like knitting, crocheting, and woodworking. Inexplicably, painting, music, and theater were not listed.

There is some practical urgency in learning more about the elderly's relationship to the arts in the real world and not solely in the laboratory. Arts programmers need to know more about the preferences of the elderly in order to design "age-friendly" events that appeal to active oldsters and give creative expression to a population whose lives are bereft of these kinds of arts activities. The aged's views about the arts are "important for program directors and practitioners for planning effective art programs" (Stone, 1991, p. 31). Community programs for the old work best when tailored to their needs, interests, and abilities. When little money and few resources are available, priorities for arts programs need to be justified. Will seniors enroll in dance or physical exercise courses? Which is more attractive: transportation to a poetry reading, a class on music history, a free ticket to an opera, or a workshop on writing a life review or memoir? Further, how does the elderly's participation in arts programs compare with other areas in which they are traditionally active, such as volunteering, golf, card playing, gardening, travel, and adult education?

Little is factually known, though, about the actual kinds of arts activities pursued by the elderly, and equally important, how these compare with younger individuals. Do more elders paint than visit art museums, and how do these rates compare with the young? In addition to determining young–old differences in behavior related to the arts, it is also important to gauge the elderly's attitudes toward and interests in the arts, and if these differ from younger individuals, it is because these sentiments are the basis for what they do in the arts.

But as is so often the case for questions about the life course, developmental studies of leisure activities in the arts focus largely on youth, concentrating on either children or college students, and to a lesser degree, the middle-aged. There is the presumption that the "recreational benefits of leisure have less relevance for older persons than for young adults" (Tinsley et al., 1987, p. 55). Little is therefore known about the activities, attitudes, and interests of the elderly in relation to the arts and if these differ from the young. This is the question addressed in this chapter.

EXPECTATIONS

The relationship between aging and the arts might take one of several forms. Following the decline models of creativity and aging (Chapters 1 and 3), the old will do less in the arts than the young because of their sensory, cognitive, and other limitations, decreased mobility and lessened strength, and reduced visual and auditory capacities. Consequently, the elderly will rarely visit art museums, attend concerts, or take in the theater. Further, they will find it difficult to play an instrument, write a story, or hold a book, or indeed, comprehend and remember what they read. The elderly's level of activities in the arts, following the decline argument, will therefore be lower than younger audiences and participants who are better equipped sensorily, more physically spry, and with greater cognitive capacities. As a consequence of the elderly's minimal engagement with the arts, their attitudes toward and interests in the arts will be less positive than the young. The old might prefer, at most, relatively passive activities, like listening to music at home, rather than attending a live music concert, and they would rather watch a program about art on TV than visit an art museum.

With participation in, attendance at, attitudes toward, and interests in the arts falling with age, the decline thesis would expect the elderly to be less familiar with, open to, and enthusiastic about arts events, and their appreciation of and sensitivity to paintings, music, and the like will be blunted and diminished, compared to younger audiences. Confirming these doleful possibilities, the elderly attend fewer arts events than any other age group with the exception of teenagers (DiMaggio & Useem, 1980; McCutcheon & Tecot, 1985; Malcahy & others, 1980).

These foreboding predictions, even if borne out, are open to alternative explanations other than aging's declines. Low attendance at arts events and minimum participation in the arts are not necessarily signs of the elderly's lack of interest, cognitive and other mental losses, or narrowed esthetic sensibilities. Instead, reduced activities could reflect the high cost of arts events for seniors living on restricted budgets, the unavailability of public transportation for the elderly who cannot drive, and frail oldsters' fears about attending evening events in downtown areas where crime rates are high. These external barriers to the aged's greater involvement in the arts, not aging's inner infirmities, may lead to a weakened attitude toward and a reduced interest in the arts. There are, in other words, several ways to explain the elderly's negative relationship to the arts, and these do not reflect cognitive, attentional, and affective shortcomings.

The relationship between age and the arts can also be a positive one. Studies of old painters and old nonartists' reactions to paintings in previous chapters testify to the power of art to bring out the best among the aged and to disclose abilities that would otherwise be ignored. Older retirees have the leisure time and financial resources to participate more fully in the arts than the young who are limited in time, opportunities, and money if they are in school, or as young adults, are constrained by restricted budgets, a growing family, a new home, and more preoccupied by careers than concerts. Occasions for leisure among the aged also compare favorably with the middle-aged, whose time, energy, and money are set aside for sports, travel, and children. The arts may therefore have a lower priority, less discretionary funds set aside, and fewer occasions to frequent among the young than the old.

A case can therefore be made for the old, rather than the young participating more in the arts, having stronger interests, and showing a more positive attitude. These predictions, which emphasize the advantages rather than the shortcomings of the old, reflect the position taken by advocates of aging programs, successful aging, and humanists who believe in gains accruing from growing old.

The elderly's engagement with the arts could also be equal to the young. The strengths and weaknesses of the young, such as faster responses, and in the case of the old, greater experience, could offset each other. Participation, attitudes, and interests in the arts for the two age groups will therefore be equivalent, as were their responses to paintings (Chapter 15). Few or no differences between the young and old's involvement with the arts would support the constancy position on aging which emphasizes the stability of abilities over time (Chapter 5).

The elderly might also demonstrate a mixed profile. With increasing age, some activities in the arts improve, while others remain the same, decline, or cease, depending on the particular sensitivities tapped by specific events. Thus, some activities in the arts might favor one age group or the other, depending on the kind of event.

In Summary. The decline position predicts losses in the elderly's participation, interests, and attitudes in the arts. The continuity view is more optimistic, expecting the old to do as well as the young. Other perspectives expect the old to do more or better than the young, or a combination of gains, losses, and equality.

QUESTIONNAIRE STUDIES: METHOD

The relationship between age and the arts, described in terms of activities, attitudes, and interests, was investigated with different questionnaires: (1) The "Arts Activities" questionnaire covered two kinds of activities: (a) attendance at arts events, like concerts and visits to arts museums; and (b) participation in arts activities, such as painting and composing music. (2) The "Importance and Challenge of the Arts" questionnaire measured attitudes toward four major areas of the arts, like music. (3) The "Interest in Four Areas of the Arts" questionnaire focused on the respondents' interest in four areas of the arts, for both the present and over the lifetimes. Specific details about these questionnaires are given later.

The three questionnaires were completed by 20-, 40-, 60-, and 80-year-olds. All but the 20-year-olds were paid volunteers who had served in previous research at a large midwestern university or answered newspaper advertisements; the 20-year-olds were volunteers from university courses who participated for credit. The number of useable returns ranged from 138 to 149, depending on the particular questionnaire; about as many men as women participated.[2] The respondents were middle-class, in good health, and relatively privileged in income and education. Thus, 63% of the respondents over 40 had at least a BA and 28% had at least an MA. Nearly half were married (44%); the remainder were single (24%), widowed (15%), divorced (11%), or living with a partner (5%). Nearly all participants (96%) were Caucasian. The sample therefore represented optimally functioning individuals, neither institutionalized in hospitals nor confined to old age homes, and with sufficient leisure time and income to take advantage of opportunities in the arts. The questionnaires were completed at the respondents' convenience and returned in person to

---

[2] The three questionnaires were part of a larger battery of questionnaires, along with laboratory tests on cognitive, personal, physical, social, and affective aspects of aging. The studies reported here were therefore a small part of a much larger project supported, among others, by AARP. Thus, the "Arts Activities" questionnaire was taken from a larger set of 70 questions about various activities; the "Importance and Challenge of the Arts" questionnaire accompanied 28 additional areas of everyday life; and the "Interest in Four Areas of the Arts" questionnaire was embedded with 26 other questions about changes over time in different areas. The questionnaires were developed from and pretested in several pilot studies. For example, the "Interest in Four Areas of the Arts" questionnaire drew on the activities chosen by a pilot group of 79 respondents, demographically similar to those who participated in the present study.

The number of respondents who completed each questionnaire were as follows: 138 answered the "Arts Activities" questionnaire; 141 returned the "Importance and Challenge of the Arts" questionnaire; and 149 responded to the "Interest in Four Areas of the Arts" questionnaire. The same number of men and women completed the first questionnaire (69 each); somewhat more women than men returned the second questionnaire (76 and 65, respectively); and men and women were fairly evenly distributed on the third questionnaire (78 and 71, respectively). Their average ages were as follows: for the 20-year-old group, Mean=24.10 (SD=3.30); for those in their 40s, Mean=43.70 (SD=2.70); for the 60-year-olds, Mean=64.3 (SD=3.00); and for respondents in their 80s, Mean=83.4 (SD=2.60). (The SD, or Standard Deviation, is a measure of range, with $+/-$ 1 SD accounting for 73% of the sample falling above and below the mean. Among the 80-year-olds, for example, nearly three quarters ranged in age, on average, from 80.80 to 85.40.) Each age group, with one exception, was about equally divided by gender; the difference between the number of men and women did not vary by more than four. At age 80, though, women predominated: the "Arts Activities" questionnaire was completed by 9 older men and 15 women; the "Importance and Challenge of the Arts" questionnaire included 10 men and 19 women; and the "Interest in Four Areas of the Arts" questionnaire was turned in by 11 men and 20 women. Gender differences in the responses to the questionnaires, though, rarely differed statistically, and when they did, were unrelated to age. Consequently, age differences in relation to gender are not discussed, although they are mentioned in the footnotes on those rare occasions when they were statistically significant.

the university's Psychology of Aging laboratory where additional tests (not reported) took place.

AGE DIFFERENCES IN ARTS ACTIVITIES

The respondents reported on the frequencies with which they attended and participated in 11 different activities in the arts over the past 5 years. The five kinds of arts events attended were theater (live drama), dance concerts, musical events (a symphony, concert, or opera), a poetry reading, and visits to art museums. Six types of participation in the arts were listed: writing a short story, novel, or play; writing a poem; playing a musical instrument; composing a song; painting a picture; and sculpturing (or working with clay). The frequencies reported for the participatory activities were highly variable and included many non-responses, distorting the comparisons. Consequences, the frequencies given by each individual were converted into an activity engaged in at least once (scored as "1") or not ("0"), and the former were reported as percentages for each age group. In addition to the listed 11 areas of attendance and participation, respondents could write in examples of "other" areas of activities.[3]

Attendance at Arts Events

Older respondents attended more arts events than younger ones. Consider attendance at the theater, where increases with age were most dramatic (Figure 16.1). Attendance grew regularly from the 20s to 80s; the mean attendance rates for the four age groups over a 5-year period, on average, increased from a low of 5.49 to a high of 15.72, flattening only slightly in the 80s.

Higher attendance with increasing age also characterized the other areas, although not as sharply or as consistently as theater. Attendance at musical concerts increased mainly between the 60s and 80s, averaging about 17 and 22 occasions, respectively. Attendance at dance

---

[3] Listed in addition to the six participatory activities were gardening, knitting/crocheting, and taking photographs. However, the frequencies reported for these three activities were much more variable than the others. For example, some respondents had taken several thousand photos, while others reported a few hundred, a dozen, or none. Consequently, the results for these highly variable activities were not reported in the text; some aspects unrelated to age, though, are mentioned in various footnotes.

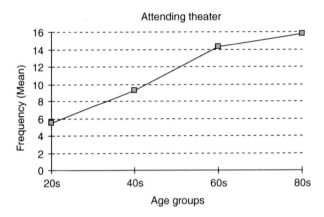

Figure 16.1. Age Differences in Attendance at the Theater.

concerts increased markedly from the 40s to 60s, from about 3 to 10, respectively, on average. The 80-year-olds' attendance at dance concerts dropped precipitously, but it was still about the same as respondents in their 40s. Attendance at poetry readings, which was very low in general, nonetheless increased dramatically (almost fivefold) from the 60s to 80s, from 0.24 to 1.14 visits, on average; the oldest age group attended at almost the same rate as those in their 40s (1.36). Of the six areas, only visits to art museums declined with age: it was highest in the 20s (at about nine visits over 5 years, on average), then fell to almost half that for respondents in their 40s and 60s (both groups attended at about the same frequency, 4.93 and 4.89, respectively), and then reached its lowest point in the 80s (at 2.75).[4]

---

[4] Statistically, age differences in attendance at the theater were significant ($F(3,576)=2.66=p<.05$), and while attendance at each of the other four arts events was not ($F(3,576) \leqslant 1.21$, $p>.05$), age differences in attending the five areas, on the whole, were suggestive ($F(12,508)=1.47$, $p<.13$). Statistical tests, here and elsewhere, that fell short of generally accepted levels of significance in the social sciences (i.e., probability of occurrence, $p<.05$), were labeled and interpreted as follows: Where the statistical probability was somewhat short of traditional cut-off points (i.e., $p<.10>.05$), these were labeled "trends." Where the pattern was in a particular direction but not statistically significant, these were termed "tendencies" or "suggestive." A fairly liberal attitude toward statistical tests was justified because the self-reports were based on memories spanning a 5-year period, involving highly subjective and subtle judgments, thereby permitting some leeway. Readers may interpret these statistics and labels differently.

Participation

Participation in the arts, too, differed with age, but unlike attendance which gained, it declined. The 20-year-olds were more active than any other age group, followed by respondents in their 40s. Declines with increasing age consistently marked the two kinds of musical activities: An instrument was played by 67%, 50%, 19%, and 3% of the 20-, 40-, 60-, and 80-year-olds, respectively. Composing a song ranged from highs of 19% and 15% for the 20- and 40-year-olds, respectively, to 0% and 3% for the 60- and 80-year-olds, respectively. Painting, too, followed a consistent pattern of decline for the four age groups: 34%, 27%, 15%, and 5% painted among the 20–80-year-olds, respectively.

But there were some notable exceptions among the two older groups. In keeping with the above pattern, 20-year-olds wrote more stories, plays, and novels (at least once in the last 5 years) than the other three groups, where the percentages in participatory activities steadily declined with age: 61%, 36%, 19%, and 24%, for the 20-, 40-, 60-, and 80-year-olds, respectively. Note, however, that the 80-year-olds did somewhat more writing than the 60-year-olds (24% and 19%, respectively). On the writing of poetry, the 20-year-olds again far outstripped the others (46% wrote poetry at least once), but the two older age groups were at almost the same level as the 40-year-olds (22%, 19%, and 26%, respectively). Twenty-year-olds also sculpted more often (28%) than the 80-year-olds (5%), but the 40- and 60-year-olds hardly differed (17% and 14%, respectively).

A broader analysis, furthermore, mutes the picture of declines in participatory arts activities with age. The six participatory arts activities were combined into three related groups: writing (a story, novel, and play, as well as a poem), music (playing and composing), and visual arts (painting and sculpting). Declines with age were now more

Continued

Irrespective of age, but of some interest, attendance in the five areas of the arts differed from one another. Attendance at musical concerts was highest, occurring on the average 17.62 times over a 5-year period. It was followed at some distance by attendance at live theater (Mean=11.21). Considerably less frequent were visits to art museums and attendance at dance concerts (Mean=5.49 and 3.81, respectively). The activity least often attended was poetry readings (Mean=0.92). The differences between these activities were statistically significant $(F(4,508)=25.24, \ p<.01)$. Attendance at dance concerts differed by gender $(F(3,576)=2.66, \ p<.05)$, a rare distinction in the studies reported here and elsewhere in the book: Women across all four age groups attended more events than men did.

suggestive than definitive (statistical probability levels were no longer significant). Thus, a more global measure of three types of participation in the arts showed fewer age differences than the more specific focus on six particular areas.[5]

In Summary. Declines occurred with age for participatory activities, like playing an instrument. The 20-year-olds did more with art than any other age group. There were some exceptions, though. The 80-year-olds wrote more fiction than the 60-year-olds; both older age groups wrote about the same amount of poetry as the 40-year-olds; and the 40- and 60-year-olds did about the same amount of sculpting. College students were probably more actively engaged in the arts because they had more opportunities to take art and creative writing classes.

However, greater campus venues for the arts did not result in college-age students attending more. In fact, just the opposite was the case: attendance at arts events generally increased with age, just the opposite of participation. Thus, relatively passive involvement in the arts, observing rather than doing art, favored the older groups.

Other Activities in the Arts

The respondents also wrote in "other" participatory activities in addition to those that were listed, such as square dancing, singing in a choir, and making greeting cards. These showed a somewhat different pattern with increasing age than the six listed. The 40-year-olds, not the 20-year-olds, were the most active. Nearly half (48%) reported an "other" artistic activity in the past 5 years. Further, the number of 60-year-olds' "other" activities were not that different from the 20-year-olds (33% and 40%, respectively). These shifts toward the older ages probably reflect the kinds of activities engaged in by middle-aged but

---

[5] Age differences in participatory activities were statistically described as follows: For the writing of stories and poems, $F(3,843)=2.99$ and 7.47, respectively, $p<.05$; for playing an instrument, $F(3,843)=17.21$, $p<.01$; for composing a song, $F(93,843)=1.71$, $p<.20$; for painting, $F(3,843)=3.78$, $p=.01$; and for sculpting, $F(3,843)=1.77$, $p<.20$. For age differences in general, $F(6,840)=2.01$, $p<.01$. Men and women at different ages did not differ in their rates of participation $(F(6,840)=0.76$, $p>.05)$ or in general $(F(1,3)=0.22$, $p>.05)$. When the six activities were combined into three similar groups, a "trend" for age differences was found $(F(3,358)=1.90$, $p=.08)$.

not college youth. The 80-year-olds, though, still lagged far behind the other age groups (19% participated in "other" activities).[6]

Conclusions

It makes sense (now that the facts are known) for active participation in the arts to decline with age, since painting and the like take more energy and require frequent travel (shopping for supplies, taking classes, going to exhibits, carrying materials, working outdoors). Similarly, it stands to reason (given the facts) that relatively passive activities remained stabile with age, since attending a musical concert and the like requires relatively less energy and time than playing an instrument (and no lessons). Thus, an arts event was attended at least once within 5 years by at least 50% of each age group while the number participating in some kind of art or another never rose above 43% at any age, including the youngest. Even among the 80-year-olds, attending an arts event occurred more frequently than participating in some arts activity (50% vs. 10%, respectively). Thus, every age group was more often at an arts event than doing some kind of art.[7]

---

[6] Participatory activities listed under "other," in addition to the examples cited in the text, included telling jokes, writing letters, spinning wool, sewing clothes, making baskets, constructing stained glass windows, building a canoe, carving wooden duck decoys, raising children, cooking meals, making salads, developing a cactus collection, and redecorating a room. The age differences for "other" creative activities were statistically significant ($F(3,843)=2.99$, $p<.05$).

Respondents also wrote-in arts activities attended, in addition to the five listed. However, these were too infrequent (and much fewer than participatory activities) to be analyzed. They listed going to exhibits of wild-life paintings, attending lectures on arts, culture, and literature, watching the ice-capades, viewing ethnic events and festivals, bird watching, listening to talks, travel, attending track and field events, and going to baseball games.

Traditionalists might consider these "other" participatory and observational examples as stretching the definition of "art" and "creative" activities. Nonetheless, they reveal the richness and appeal of the popular arts among the general public.

[7] Participatory activities in the arts were much more variable than attendance at arts events. In order to make the two sets of numbers comparable, attendance figures were converted into the percentage of activities attended at least once in the past 5 years. For example, a respondent who checked off one of the seven participatory activities (including "other") and attended two of the six arts events was scored as 14% and 33%, respectively. ("Other" arts events attended were not counted in this analysis since they were too few to be meaningful additions; the set of three activities headed by "gardening," omitted in the main analyses (see footnote 3), was included instead.

Both age differences and similarities therefore described attendance at and participation in arts activities, depending on whether the event was relatively more active or passive, and within each, the specific kind of art referred to. The inclusion of multiple measures (participatory and attending, listing 11 specific activities and "other") was critical in differentiating arts behavior at different ages. Thus, when all five areas of attendance were combined, differences between the four age groups' attendance became less marked: the frequencies reported by 40–80-year-olds ranged from 50% to 55%, not much less than the 60% rate for the 20-year-olds. As already noted for the combined participatory activities, when the six were combined into three larger groupings, declines were less pronounced. But when both participatory and attendance activities were combined, only decline was evident: the rates for the 20–80-year-olds fell from 56% to 47%, 43%, and 33%, respectively. Thus, had the respondents been asked only one question, such as "Indicate the extent of your involvement in any kind of arts," age differences might have been indeterminate, absent, blurred, reversed, or exaggerated. On the other hand, a more inclusive questionnaire with a greater number of activities associated with older respondents, such as crocheting, needlework, sewing, glasswork, beadwork, and woodworking, would have been advantageous to the old, while the young would have gained if rock concerts and "happenings" were included.

AGE DIFFERENCES IN ATTITUDES TO THE ARTS

The same four age groups rated the importance of four areas of the arts to (1) their "well being, happiness, and life satisfaction" and (2) as a "mental challenge." The four areas referred to writing (letters, poems, stories, diaries, and reports), reading, arts and crafts, and music (playing an instrument, singing, or composing a song). Since the two sets of ratings for importance and challenge did not statistically differ by age,

Continued

The statistics for the combined participatory and attendance data were as follows: Participation in the arts declined more with age than attendance at arts events ($F(6,280)=2.72$, $p<.05$). Across all ages, the arts were more often attended than performed ($F(2,280)=82.94$, $p<.01$). Women generally did more of both types of activities than men (Mean$=47\%$ and 42%, respectively), a distinction found at every age ($F(1,3)=3.84$, $p=.05$). There was no age by gender interaction ($F(3,140)<1$).

they were combined into a single score that indicated the respondents' general attitude toward each of the four areas of the arts.[8]

The older respondents' attitudes toward the arts were generally more positive or about the same as the younger groups. Thus, the 80-year-olds' attitude to both writing and reading was higher (means were 3.25 and 4.20, respectively) than the three other age groups (mean ratings ranged from 2.89 to 3.01 for writing and 3.60 to 3.73 for reading). The 80-year-olds' attitude toward music, furthermore, was nearly identical to the 40-year-olds (2.49 and 2.46, respectively). The oldest group, however, had the least favorable attitude toward arts and crafts of all age groups (1.91 vs. 2.19 to 2.34, respectively), an area generally rated low by all age groups. The attitudes of the 60-year-olds, furthermore, were similar to the two younger groups on writing, reading, and arts and crafts. Other than the differences noted, the attitudes of the four age groups toward four areas of the arts were similar to one another.

In Summary. Attitudes toward the arts, like attending the arts, generally favored the old, unlike participating in the arts, which gave the advantage to the young, although there were exceptions. The relationship between art and age is therefore mixed. What is clear, though, is the old are not at a disadvantage in a variety of arts-related areas, just as was true when responding to paintings.[9] Would this also be the case for their interest in the arts, the last questionnaire?

AGE DIFFERENCES IN INTEREST IN THE ARTS

The four age groups rated their interest in the arts in general, for the present and over their lives. For comparison purposes, they also rated their current and long-term creativity. (Further perspectives were

---

[8] The importance and challenge of each of the four areas of the arts were rated with a 5-point scale. A rating of "1" indicated that an activity was "not at all important [or] challenging," "3" meant it was "fairly important [or] challenging," and "5" indicated that the art form was "extremely important [or] challenging." The importance and challenge ratings did not differ from one another as a function of age $(F(3,133)<1)$, thereby permitting the two scores to be combined into a single value that reflected the respondents' general attitude toward the arts.

[9] Attitudes to the arts tended to differ by age across all areas of the arts $(F(9,399)=1.78, p<.07)$. Specific age differences were as follows: music, $F(3,499)=3.08, p<.05$; reading, $F(3,499)=2.26, p=.08$; art and crafts and writing (both at $F(3,499)\leqslant10, p>.05$). Attitudes toward these four areas differed from one another, irrespective of age $(F(3,399)=88.59, p<.01)$, but there was one gender difference: 80-year-old women $(F(1,133)=16.08, p<.01)$.

provided by the respondents' ratings of their long-term interest in spiritual matters and wisdom, reported in footnote 10.)

Interest in the arts increased with age, another positive entry for the elderly. From the 20s to 40s, growth was modest, with ratings at 3.23 and 3.39, respectively, but interest among the 60- and 80-year-olds sharply increased, with means of 3.58 and 3.66, respectively. In contrast, creativity's ratings for the 20- to 60-year-olds' creativity were fairly similar (ranging from means of 3.43 to 3.57), although they were lower for the 80-year-olds (at 3.15). The latter is the only instance where creativity was rated lower than an interest in the arts; the three younger age groups rated their creativity as either higher or the same as their interest in the arts.

This pattern changed markedly when the respondents rated their interest in the arts over their lifetimes. Unlike current interests, which increased with age, the profile was essentially flat, although still quite high; mean ratings ranged from 3.64 to 3.86. Long-term interests in the arts for all age groups compared to lifetime ratings of creativity were now consistently higher (the latter ranged from 3.23 to 3.41, respectively).[11]

---

[10] The rating scale for the respondents' interest in the arts (both current and long-term) ranged from "very low" (=1) through "average" (=3) to "very high" (=5). Ratings of their interest in the arts over a lifetime ranged from "greatly declined" (=1), to "remained the same" (=3) to "greatly increased" (=5). The same values were used to rate creativity, spirituality, and wisdom.

[11] The statistics for interest in the arts were as follows: Current interest in the arts and self-ratings of creativity tended to differ by age ($F(3,142)=2.52$, $p=.06$), as was true for long-term interest in both of these areas, as well as self-ratings of wisdom and spirituality ($F(9,420)=1.66$, $p=.10$); the latter, not discussed in the text because they detract from the main findings, were added to provide a broader context.

Irrespective of age, the four sets of ratings over the long term differed from one another. Interest in the arts received the second highest rating, just behind wisdom (Mean = 3.73 and 3.92, respectively), which received the highest rating ($F(3,420) = 16.31$, $p<.01$). Interest in the arts was rated higher than a concern with spiritual matters (Mean=3.54) and an interest in creativity (Mean=3.33), which placed third and fourth, respectively (t-tests, p<.05).

Gender differences did not differentiate the four age groups' interest in the arts or the other three areas ($F(1,142)=0.88$, $p>.05$). With the four areas combined, gender and age tended to be factors ($F(9,420)=1.92$, $p\geqslant.06$). As in most analyses, gender distinctions in the arts were largely irrelevant.

The long-term ratings of spirituality and wisdom were more variable than interest in the arts. Of particular note, wisdom received the highest ratings from each age group (means ranged from 3.57 to 4.28), but surprisingly, it was not the two older groups who gave it the highest scores. This unexpected finding was also true for spirituality, with younger individuals rating it higher than the 60- and 80-year-olds. Perhaps older individuals were more realistic and less idealistic about wisdom and spirituality than younger respondents.

In Summary. The two older groups' current interest in the arts were higher than the two younger groups, and more like their attendance and attitudes than participation. This pattern changed when interest was measured over the long run but not to the disadvantage of the old: all age groups were equally interested in the arts. However, since the ratings were so high, there was little opportunity for differences with age to appear. Nonetheless, whether the time period is short or long, the elderly's interest in the arts compared favorably with the young and with their self-reported creativity.

CONCLUSION

A mixed and uneven pattern of gains, losses, and equality with age described the relationship between the arts and age. The young were generally more active participants in the visual arts, music, and writing—but not always; and the elderly's attendance at arts events were often better or the same as the young, or sometimes worse, depending on the type of event and whether old was defined as the 60s or 80s. A varied pattern also characterized different age's attitudes to four areas of the arts and interest in art generally, past and present: the young had more positive attitudes and greater interests than the old in some cases, the opposite was true in other instances, and they were occasionally about the same. What does seem safe to say is that the 60- and 80-year-olds' participation in, attendance at, attitudes toward, and interests in the arts were not seriously impaired. This is a remarkable finding in light of the physical, sensory, and psychological demands made by the arts on aging audiences and participants, and the complex nature of the various kinds of art and the skills they require. To the extent the young were superior, it was mainly in participatory activities, albeit not uniformly. Thus, the elderly were less active in painting, playing an instrument, and composing, but not in writing stories, poetry, or sculpting. Even with reduced participation in some areas, the elderly's attitudes toward and interest in the arts, compared to younger respondents, were not adversely affected, just as they were not at a loss in responding to paintings (Chapter 15).

The sensory declines, physical losses, cognitive handicaps, and other debilities of the aged did not appreciably hamper their ability to engage in various forms of artistic endeavors or interfere with their attitudes toward or interest in specific areas of the arts or the arts generally. Perhaps gains in knowledge, maturity, and experience (Part IV) made it possible for them to find ways to compensate for and adjust to losses, thereby making them equal to younger audiences in some areas of the arts, if not superior in some cases. Age-related shortcomings were minor, irrelevant, overcome, or translated into advantages.

For men and women with adequate leisure time, sufficient income, fair health, adequate mobility, and a good education, the presumed handicaps of aging, in so far as they apply to activities, attitudes, and interests in the arts, were largely immaterial or adequately managed. For most respondents in their 60s and 70s, and for many in their 80s, the expected losses of old age did not impair a wide swath of arts-related beliefs, feelings, and behaviors; losses were few or nonexistent, and in a number of instances gains were the rule. For the elderly to remain active, engaged, and committed to the arts, as these studies indicated, they must have many cognitive, affective, personal, and behavioral competencies. Although the old were not always as good as or better than the young, especially in more active forms of artistic activities, the overall profile is consistent with the positive picture drawn of aging artists' creativity (Parts III and IV).

The generally positive relationship between aging and the arts, for both artists and nonartists, justifies an optimistic perspective on old age. Abilities may not weaken as much after age 60 as many believe, and creativity need not be lost, as some fear. Old age need not be a time of precipitous declines, unmanageable handicaps, and far-reaching restrictions. There is no reason to believe that the capabilities of the aged are limited to the arts, but also hold for similarly complex and demanding areas outside of the arts.

The long-term stability, gains, and minimal declines in the elderly's response to paintings and the arts in general compares favorably with the young (this Part) and old artists' late-life creativity (Parts III and IV), and speaks to the persistence of "good" cognitive, personal, and other traits and abilities into old age. Sustained achievements support a continuity model of creativity, point to the insufficiency of the decline position, augment the possible ways of successful aging, and demonstrate the value of art to everyday life. Keep in mind, though, that just as the aged do not do well in all aspects of the arts (slower reaction times, less active participation) there is considerable evidence for losses occurring with age in creativity, cognition, and other abilities (Chapter 3). Yes, there are declines with age, but not in everything, and there are gains, too.

These optimistic conclusions follow from research with art. For most people, though, whether elderly or not, the relevance of the arts to everyday activities and the vitality of aging are neither recognized nor appreciated, or seen as legitimate topics for scientific inquiry.

The pertinence of the arts to the well-being of the old, psychological theories of late-life creativity, models of aging in gerontology, and the methodology of science, are discussed in the concluding chapter, along with the role of the arts in illuminating broader concerns about aging, creativity, and cognition.

CHAPTER 17

Looking Ahead

Don't complain about old age. How much good it has brought me
that was unexpected and beautiful. I concluded from that that the
end of old age and of life will be just as unexpectedly beautiful.
            Tolstoy, Letter to V. V. Stasov, 1906 (written at age 78)

To recapitulate: Studies of the late-life creativity of aging artists,
old art, and the elderly's responses to paintings and engagement in a
variety of arts activities yielded a highly positive picture of growing
old. Creative efforts by aging artists continued, creative expression took
new and unexpected turns, old creative artists gave glowing accounts
of their late-life abilities, the creativity of old artists neither peaked as
early nor fell as precipitously as expected by proponents of early
decline, and artistic creativity spiked more than once after youth. The
course of artistic creativity was therefore considerably extended into
old age, at least for painters. The profile for aging non-artists was
similarly encouraging. They responded to paintings much as younger
viewers did, and their activities, interests, and attitudes toward the arts
were similar to younger audiences, sometimes better, and only occasionally worse. The aged were not overwhelmed by the perplexities
of art.

The quality of aging painters' work, and the parity between young
and old arts audiences, are welcome auguries of late-life's possibilities.
Growing old was not necessarily a time of loss but an age when mental abilities were maintained as well as invigorated. The inevitable
handicaps of old age did not dampen creativity or blunt artistic sensitivities. These affirmative findings counter a limited perspective on old
age, support the continuity of creativity, broaden the notion on successful aging, and promote the place of the arts in the study of late-life
development in psychology and gerontology. The complexities of art
showcased the talents of aging artists and old people in general. Yet
when tested with ordinary laboratory materials, the elderly do worse
than the young, rarely the same, and hardly ever better (Chapter 15).
Therein lies a paradox.

The old do well in studies that include art, but not so well
with line drawings, geometrical patterns, disconnected forms, and
nonsense shapes in the sensory–visual–perception laboratory, or with

meaningless words, isolated text, and out-of-context prose in verbal learning and remembering experiments. When tested with the limited and limiting materials of scientific investigations, and faced with procedures that require a fast, single, and correct response, the abilities of the elderly are either underestimated or overlooked (Chapter 15). Circumscribed conditions are especially troublesome for healthy, competent, and independent older research participants who are accustomed to freedom in their occupational and other roles at home, in the neighborhood, at recreational activities, and with voluntary organizations. In a controlled setting, though, autonomous elders are constrained from asking probing questions or replying with extended answers. Understandably, they are likely to become impatient with meaningless tasks, uneasy over unexplained procedures, frustrated by truncated explanations, skeptical about their ability to master meaningless material, confused over stimuli without a context, and resentful of youthful supervisors who, by the nature of scientific inquiry are strict, distant, and uncompromising. Thus, for reasons only indirectly and tangentially related to age, elderly but capable research participants are unable to perform as well as they do outside of the laboratory, and hence they do worse than their youthful counterparts and not much better than their institutionalized brethren.

The restrictive conditions that put the elderly at a disadvantage have been recognized by others and corrections were recommended, such as putting more emphasis on problem-finding and questioning, qualitative thinking, empathizing, and synthesizing, on which the old can excel (Craik & Simon, 1980; Labouvie-Vief, 1980; Salthouse, 1989; Wass & Olejnik, 1983). The substitution of art, including literature, for prosaic visual and prose materials, though, has not been recommended. Why not?

ART IN THE LABORATORY

Art is messy, carrying a melange of the relevant and irrelevant. What really matters in a painting and what does not? Colors or shapes? Meanings or feelings? Art evokes a multitude of sights and sounds, words and feelings, judgments and prejudices, and these are difficult to separate. Which of the many reactions to a painting are critical? Shapes, colors, composition, light, shadow, space? All of the above? More than the above? Which aspects of what part of how many of the multiple components of music, literature, theater, dance, and sculpture lead to what sorts of consequences? Art bombards and overwhelms its audiences—as is often its intent—with sensations, perceptions, memories, meanings, and interpretations. These change within seconds and vary from person to person, depending on their previous exposure to art, as well as education, social background, and cultural values. Even so mundane a task as choosing a work that is liked more than some other one is a multi-layered event fraught with extraordinary subjectivity. Furthermore, feelings about art are often unverbalized and unverbalizable, unspecified and unspecifiable, not to speak of unpredictable, and includes a great deal of deeply personal feelings and unconsciously held beliefs difficult if not impossible to communicate. Responses to art, moreover, are often stereotyped, expressed in simple and uninformative ways ("it's cool," "neat") lifted unthinkingly from popular culture. Alternatively, some reactions to art can be original, spontaneous, revelatory, heartfelt, thoughtful, and insightful.

Whatever is grasped is then subject to the simplifications, controls, and statistical funnels of scientific study that mute if not nullify the vibrant, dynamic, unique, and transcendental qualities of art that artists, scholars, and lay audiences consider essential. No wonder that scientists, if they turn to experts for guidance on what to study and how to proceed, receive little except skepticism or a cryptic response (Lindauer, 1984a).

Art carries another burden: its atypicality. Looking at or doing art is not a common activity for most people; the arts are not a part of everyday life but an occasional activity for a relatively small number. Art is, after all, an illusion, and some might say in the case of modern art, quite unreal. The arts are narrow and specialized areas of activity that require a modicum of education, experience, and training.

A case can be made, though, for including art in research on aging (Chapter 15). Art indulges the aged's propensity for "story telling," that is, to elaborate and explain. Faced with the dull materials of the ordinary laboratory, elderly participants have little reason to tell a "story" about truncated and innocuous visual and verbal stimuli, even if encouraged to do so. Realistic, historical, religious, and other kinds of representational art, though, have a "story." Less obviously, so does abstract, impressionist, expressionist, avant-garde, and other kinds of modern art. A viewer's curiosity is piqued, if not perplexed, shock and surprise are elicited, and perhaps outrage. "This is art! My grandchild can do better!" Nonetheless, older viewers will courageously, if asked, try to make sense of unrecognizable forms, suggestive shapes, and splashes of random colors by inventing a "story," including speculations about the artist's intent, ability, and "sense," if not sanity.

The elderly's propensity to tell stories is enhanced by a holistic search for meaning, also associated with old age. The elderly seek the essence of a work, the "big picture," its "gist" (Commons, Richards, & Armon, 1984, especially pp. xiii–xxviii). (For the connection between "getting the gist" and post-formal thinking, a type of reflection that occurs after late adolescence and adulthood, see Sinnott, 1998. Discussions of the post-formal stage, however, rarely go beyond age 65 or place it in the context of creativity or the arts.) Young perceivers, in contrast to older ones, are predisposed to look for details and be accurate. Consequently, their scanning strategies are specific and thorough, geared to extracting, as efficiently as possible, a quick, correct, and complete response. Focusing on details works well with isolated and fragmentary visual and verbal arrays with little meaning, the traditional stuff of the laboratory. Complex objects like art, though, call for a broader and more integrative approach that plays to the elderly's holistic and story-telling tendencies.

Art maximizes other age-related skills (Chapter 15). The elderly have amassed a considerable amount of information, or "crystallized intelligence" (Horn & Cattell, 1996), which helps them interpret art's multiple and ambiguous meanings. In contrast, the "fluid intelligence" of the young, which aids in the swift perusal of an object, is not an advantage in mulling over the subtleties of art. The elderly also tend to be problem-*finders*, prompted by the vagaries of art, while the young excel at problem-*solving*, which suits tasks with definitive answers—unlike art. With age, too, quantitative thinking shifts to qualitative thinking, a dwelling on "why" rather than "how much," which makes sense with art. There is also an increase in empathic thinking, and a greater striving for unifying and synthesizing, qualities appropriate for art (Johnson, 1990).

The inclusion of art in research captures and holds the interests of participants, and prompts a profusion of ideas and feelings. These virtues are welcomed by all ages, but they help the old more by "fleshing out a story." For older participants, too, fast and accurate responses are unsuitable with art because speed and correctness work against thoughtful responses that are neither right nor wrong, good or bad. Taking one's time and airing out opinions are not liabilities with art.

Art therefore frees older research participants from the shackles of simple and meaningless materials that magnify aging's deficiencies. Unlike a skein of scattered lines, a blur of irregular shapes, an array of arbitrary words, or a deluge of random syllables—characteristic of traditional studies of sensory processes, perception, learning, remembering, and thinking—art offers elderly respondents a richness of possibilities and a diversity of alternatives. They therefore have some latitude in maneuvering around age-related impairments. The richness of paintings and literary prose makes it easier for older learners to learn (encode) and remember (retrieve), compared to banal materials, thereby improving their performance (Park, Royal, Dudley, & Morrell, 1988; Smith, Park, Cherry, & Berkovsky, 1990). Artistic materials therefore play to the elderly's strengths, not their weaknesses, allowing them to do their best. Abilities therefore emerge that would otherwise be submerged by the burden of limited materials. Under the liberal and liberating conditions of art, the constraints of the laboratory become less burdensome. Aging participants can freely reflect, interpret, weigh alternatives, pursue ideas, select from among multiple possibilities, entertain a variety of images, and experience an outpouring of emotions. Feeling more at ease, rather than cramped or bored by the materials, the old can justifiably believe that their reactions to art are as good as younger respondents, if not better. With art, personal opinions are not only encouraged, but are considered as acceptable as anyone else's. For these reasons, the old responded well to paintings (Chapter 15) and to the arts generally (Chapter 16).

Herein lies the solution to the paradox. The poor research performance of the old in traditional studies of aging depends not so much on their age as on the kinds of materials employed. Why, then, has not more use been made of art in research on the aged?

For the same methodological reasons art is not used in research with any age group. The arts may play to the strengths of seniors, but they also contain pitfalls that hold researchers at bay. Hence, a preference for traditional laboratory materials without aesthetic value or literary panache (Lindauer, 1973). Investigations of audition utilize bare tones, not the opening chords of a Beethoven symphony; and studies of touch depend on varying grades of sandpaper swatches, not a bronze replica by Moore. Art is banned from the lab.

The elderly's reactions to art, reported in Chapter 15, illustrate the troublesome nature of art. Recall that old participants had a great deal to say, so much that the studies could not be completed in the time available. Consequently, the original plan, in which participants were to be asked to elaborate on the reasons for their responses, using "talk-aloud" and probing procedures, was abandoned. Nevertheless, older participants insisted on explaining their responses, even though unasked for. The younger participants, in contrast, said no more than required. Since the rules of research prohibit an investigator from treating groups differently, the younger subjects could not be prompted to explain the reasons for their choices, an option not available to the older participants, at least not explicitly, although they did so anyway. Thus, it was not possible to compare the unrequested musings of the older respondents with the unexpressed thoughts of the younger ones. The older group's loquaciousness also made it difficult to complete the third study (on the perception of style differences in artist's works at different ages) within the allotted time. As a result, the older group was not shown as many examples of art and artists as the younger group. With less data from older subjects, age comparisons had to be tentative and statistical analyses limited.

Scientific investigations of art, understandingly, are scarce. Researchers are reluctant to replace geometric patterns with Mondrian canvases of irregular squares bounded by colored lines or substitute suggestive Picasso sketches for line drawings. Empirical inquiry insists, and rightly so, on rigorous controls over what is included and excluded, the ability to manipulate materials and measure responses, and the predictability of outcomes. Hence, simple visual, verbal, auditory, tactile, and motoric stimuli are relied upon, not paintings, literature, music, sculpture, and dance.

Since the materials used in the studies were art, there are therefore good reasons to be cautious about the results. There are additional reasons to be tentative: the unusual phenomena (style) and atypical activities (art-going/making); the relatively small number of old artists, old works, and elderly living artists; and populations rarely recruited (artists and the healthy elderly). Moreover, the artists represented a variety of nationalities, types of art, time periods, and societies with a range of attitudes toward old age, art, and artists. Some lived in a "golden age" of creativity and others struggled without the basic necessities of life taken for granted today. The origins and circumstances of their late-life creativity were therefore uncertain. Further, not all kinds of artists and art were included. Omitted were lesser old artists who never painted a masterpiece and did third-rank work—but are nonetheless displayed in museums, sold by auction houses, and treasured by owners. Overlooked, too, were folk and non-Western art by old artists. The research was circumscribed, too, by its cross-sectional design: 20-year olds' activities in the arts were not followed into their 40s, 60s and 80s, as longitudinal studies do. Instead, comparisons were made between different age groups. The number of original studies, while appreciable (almost 20), was tiny when compared to the countless investigations of non-artists and non-art. Other unusual characteristics of the studies, including the highly specialized nature of the subject (late-life art) and the atypical participants (old artists, old arts audiences), limit generalizations about late-life development.

These objections are tempered by the recognition that all research can be criticized for not doing enough, or enough of what could be done or needed. Give it time, is a reasonable rejoinder to critics. All research, whether the participants are rats or undergraduates, and the materials are either artificial laboratory stimuli or Rorschach ink-blots, can be faulted for their narrowness and irrelevance. The adequacy of knowledge on any topic at a given time, even when pursued by the "best" science, is rarely sufficient. Fortunately, scientific research is open to improvement. Facts are never taken for granted, conclusions are not accepted by fiat, and interpretations are deservedly greeted with healthy skepticism. Scientific findings are not dogma; closed systems are anathema to the empirical spirit. But with time, as research continues and improves, doubts are reduced although probably never completely resolved.

Some objections to giving art a place in research can be met by relying on forms that are relatively unfamiliar and meaningless, thereby making them equivalent to "normal" laboratory materials, and like them, subject to controls, measurement, and the like. Indeed, for many, contemporary art is as nonsensical as anything shown in an experimental study. Studies of the perception of the elderly, for example, could display a simple but elegant Mondrian painting with its irregular squares, rectangles, and lines. Like the usual laboratory array of visual fragments, a Mondrian is like a nonsense shape. Similarly, gerontological studies of learning and remembering could present brief Haiku poetry, which approximates a disconnected skein of ordinary prose. Imagine, too, a laboratory study of figure-ground perception with an ambiguous Escher work rather than a line drawing of the oft-used reversible duck/rabbit outline; or a study of learning rates for one-line quotes from Shakespeare instead of a string of nonsense syllables. (Studies of physiognomic-expressive perception (Lindauer 1984c) illustrate the imposition of methodological controls on abstract art (Lindauer, 1970b, 1984d, 1987a) and brief titles of [Hungarian] short stories (Lindauer, 1987b, 1988; see also Lindauer, 1986a, 1990, 1991a, b). Aesthetic responses (but not necessarily aesthetic stimuli) have been scientifically studied in experimental aesthetics for over 100 years (Lindauer, 1973, 1984b).

The inclusion of the arts in scientific research, were it to increase significantly, will no doubt yield findings and conclusions that conflict with the bulk of investigations with non-art. These inconsistencies are not a cause for despair, a proof of the intractability of art, a reason to exclude complex materials from research, an excuse to throw up one's hands in defeat, or a justification for leaving the task to others for some future time. Instead, conflicting findings from the laboratory, art museum, concert hall, and theater are a challenge, an impetus for further research, and a prod for harder thinking. Reactions to art that differs from non-art open up new lines of inquiry and lead to fruitful discussions of what is gained or lost when attractive stimuli are added to the scientific mix. How do auditory detection thresholds for tones, for example, differ for the opening four notes of a Beethoven symphony? Do tactile explorations of a bronze replica by Henry Moore, compared to sandpaper swatches, reveal the unexpected sensitivities of the hand (Lindauer, 1986b; Lindauer, Stergiou, & Penn, 1986)?

Studies that include art, like those with banal materials, have to be replicated, open to criticism, and subject to improvement. The laboratory study of art requires more care and caution than ordinary materials, and empirical standards might have to be loosened with attendant losses in controls and rigor. But any methodological softness inherent in art is offset by its many advantages.

Investigations with art reported in this book have their shortcomings (postcard-sized art was shown rather than originals). Overall, though, they consistently point to one overarching fact: creative, cognitive, and other competencies persisted into old age. Whatever the flaws of a particular study, the research as a whole arrived at an affirmative view of aging, one not often found with ordinary materials. Art as a research tool has its difficulties, true, but its benefits should be recognized, also, especially when doing research on aging. If only the limitations of art are emphasized, investigations that might overcome them will never be performed and important knowledge will be lost.

The issue then comes down to this: How exacting, ironclad, easy to do, and statistically precise must empirical research be when it incorporates art? How many compromises will researchers have to make when relying on paintings, literature, music, and sculpture? This book demonstrates that including art in research on aging was not fatal. Useful information about aging would also be gained if old artists were recruited as research participants.

OLD ARTISTS IN THE LABORATORY

I have argued that art may profitably be added to the array of materials that investigate the abilities of the old. In like manner, aging artists can be included as research participants when the materials are not art. Aging artists demonstrate extraordinary abilities despite (or because of) their age (Parts III and IV). "[P]roductive inquiry could arise only from a genuine partnership between working artists and seriously informed investigators" (Koch, quoted in Franklin, 2001, p. 418). Artists are good candidates as research subjects. Imagine writers memorizing lists of nonsense syllables or painters searching for hidden figures in a confusing array of geometric patterns, two typical tasks found in learning and perception laboratories. The old artists should do well. After all, they have been solving complex problems in art and aesthetics all their lives, coping with and adjusting to aging's impact on their work when they grew older, modifying their thinking and behavior because of age-related handicaps, and producing exemplary works late in life. Aging artists have surely demonstrated the requisite skills needed to handle the prosaic stimuli of the laboratory—assuming, of course, they would volunteer to participate in studies with unimaginative materials and ordinary tasks. But if they did, old poets should be able to manage fragmented, isolated, and arbitrary strings of verbal materials; aging painters should be able to quickly grasp visual puzzles; elderly musicians should do well in differentiating machine-generated neutral tones. Old artists constantly make difficult decisions about visual, verbal, auditory, and other materials and this capacity should transfer to the laboratory. Consequently, they should do better than retired nonartists and callow youthful participants when faced with tightly controlled, time restricted, and unfamiliar materials. Elderly artists are therefore an optimal population in whom to study age-related gains and losses.

Another reason for including aging artists in research is that not many non-artists continue to work and are successful at it in old age. Most stop working around age 65, opportunities to continue life's work cease, and there are shifts to less challenging activities, like travel and golf. But old people who are still working, healthy, and competent are rarely chosen for research. Instead, more convenient populations are recruited: the retired (unemployed), institutionalized, and ill. Rarely recruited are old individuals who have a lifelong and continuing commitment to their work—like artists.

Various kinds of old artists could be asked to participate in research, but the priority should be painters. Many aging writers, composers, dancers, and sculptors have long and accomplished working careers, but only the late-life capabilities and accomplishments of painters have been abundantly validated (Part III). Among painters, furthermore, those with an old-age style (Part IV) are preferred. They have an ability to change perspectives, reconceptualize problems, redirect goals, and rechannel ideas, a flexibility of ideas and behavior that will hold them in good stead in laboratory studies.

Herein lies the solution to the paradox introduced at the beginning of this chapter: Include art and old artists in the laboratory. A more positive picture of aging should be the result. But like art, old artists are largely excluded from aging research. Pressure to bring them in, along with the inclusion of artistic materials, could come from forces outside the scientific community, from public, private, and government bodies who work with the elderly and the arts.

THE ARTS OUTSIDE OF THE LABORATORY

Policy-making bodies at national, state, and local levels promote arts programs and legislation for the elderly (Moody, 1982b, especially chapter 12, pp. 237–260; see also footnote 38, p. 300). New York City's "senior citizen cultural revolution" initiated an annual "Golden Age Exhibition" (Jankovitz, 1977). This nationwide showcase for artists aged 55 and older was viewed by more than one million people. Funded, too, was a Senior Concert Orchestra of retired professional musicians giving free concerts, a free theater ticket program for the aged, free photography and sculpture courses, an experimental theater, and a choral group. Museums, too, arrange traveling exhibitions of works by older artists to senior homes and the homebound elderly. Even department stores offer classes and lectures on the arts and assist the elderly in attending cultural events (Hoffman, Greenberg, & Fitzner, 1980; Hubbard, 1983; Training of Service Providers in Establishing Arts and Humanities Programs for Old Adults, 1981–84). There are also arts programs for the disabled elderly (Foster, 1992).

Partnerships between artists and providers of services to the aged are encouraged and facilitated by the Center on Arts and Aging, a government agency established by the National Council on Aging in 1973, which acts as a clearing house for information on the arts. "Art programs can realize the broader goals of senior centers by stimulating creativity in the aging [and towards that end] center directors... provide environments in which the creativity of the elderly can flourish" (Sunderland, 1978, p. 27). Associations between arts agencies for the elderly and the government (Famishetti, 1988) are quite extensive. McCutcheon and Tecot (1985; see also McCutcheon, 1965) listed at least 60 national organizations that offer technical assistance, seminars, and training. Programs range from creative writing, dance, drama, storytelling, and the visual arts to multimedia projects. The agenda of one conference on arts and aging strongly advocated, extensively documented, and concretely illustrated arts programs and services for older

Americans as both participants and audiences in order to "foster the meaningful and positive role the arts [play] in the lives of older Americans" (*Education: An arts/aging answer*, 1979; quote is from the abstract). Guidelines also exist for developing educational programs in the arts for the elderly (Manheimer, 1984).

Arts programs for the aged rest on the premise that the arts improve the quality of life (Chapter 16), contributing to its richness, intensity, and fullness, among other benefits.

> Practice in the arts helps the elderly to resolve conflicts, clarify thoughts and feelings, create a sense of inner order and of control over the outer world, transform negative experiences, communicate with others…sustain personal interest [and] realize the intrinsic value of old age: the harvesting and sharing of a lifetime of experience. (Sunderland, 1982; quote is from the abstract; see also Sunderland, 1976)

The arts also make the "flow" of time easier and less effortful for seniors (Csikszentmihalyi, 1997). Engaged in a satisfying activity, the elderly become unmindful of the passage of time. The belief in the importance of the arts to health and well-being is longstanding.

> Even before written history…people used dance, poetry, music, and visual art to help the sick heal and to keep the healthy strong [and that] until 250 years ago, health and art were not separate at all…. A physician's education…incorporated history, literature, music, and art. Music as well as literary and visual beauty were considered to have restorative powers almost equal to those of medicine…. Art is necessary for a productive healthy life…. Creative expression can keep people healthy. (Katz in Behrens, 1990, pp. 46–47)

More specifically, the arts stimulate the senses and maintain the elderly's "vital involvement."

> For the aging, participation in [and] expressions of artistic forms can be a welcome source of vital involvement and exhilaration [and] a setting in which every sense can be stimulated…keeping elders in touch with the natural world and their whole sensory being. The aging individual needs the satisfaction of feeling that…sensory antennae…are doing their utmost to direct and empower the aging body to remain actively involved, as well as [be a] creative agent in the world of people and materials. (Erikson, Erikson, & Kivnik, 1986, pp. 168, 188)

To accomplish these goals, the authors recommended that government policies and programs promote the elderly's participation in drama, the graphic and verbal arts (e.g., story telling), and visits to art museums, where they can also serve as docents and on boards.

The advantages of art for sensory maintenance among the aged was advanced by J. Erikson (1988), who argued that engaging in artistic events and having aesthetic experiences increased the use of the senses, thereby offsetting normal declines. In a section entitled "The vitalizing properties of creative activities" (pp. 46–73), Erikson makes the case for the arts' ability to refresh our senses by supplementing, sustaining, and enhancing sight, hearing, balance, and tactile sensitivity. Exposure to the arts, she wrote, helps the elderly appreciate more of what they once sensed more fully.

More generally, Richards (1999a) contended that creative expression not only "tends to work in the service of health [but also] helps one cope, increases physical and psychological health and well-being, and even furthers one's self-actualization and caring contribution to the world" (p. 684). Another possible consequence, she added, was becoming more "mindful," that is, better able to weigh options, be more flexible, and achieve a greater openness to cathartic and other experiences.

Perhaps because of these many benefits, Moody (1982a) reported an upsurge of interest in the arts among the old. Cultural and arts programs, he wrote, are "popular [and have] a solid niche in the aging network…. This explosion of late-life involvement in the arts by older people is a promising sign for the future [and] could point the way to a new quality of life for an aging society" (pp. 244–246).

Arts activities for the aged are not only pleasurable and satisfying in their own right, and foster well-being, but also develop competencies that improve self-confidence. The elderly, increasingly limited in so many areas of life, learn what they can do and how well they can do it. In addition to enhancing the elderly's "self-efficacy," the belief that obstacles can be overcome and problems solved, accomplishments in the arts foster the feeling that one's efforts have value and are appreciated.

Working with art also develops habits of concentration, patience, and persistence, maintains visual-motor coordination and flexibility, inspires new ideas, stretches the mind and imagination, leads to new conceptualizations, and fosters problem-meeting and -solving strategies that overcome the challenges of growing old (Runco & Dow, 1999; see also Mayer, 1999). Motivation, too, is energized, spirits are revived, and energy is revitalized. Doing art in group and busy settings helps the elderly learn new ways of coping with ambiguous, noisy, and distracting environments. In a social setting with other old artists, furthermore, friendships are formed, interpersonal relationships practiced, and networks developed. Producing art, in addition to generating a greater sense of control and freedom, provides new sources of fun and occasions for joy. "The quality of an individual's life can be enhanced through exposure to the arts regardless of age or disability" (DiGiammarino, Hanlon, Kassing, & Libman, 1992, p. 39). At the very least, involvement in the arts means staying physically active.

Skills, adjustments, feelings, attitudes, and behaviors developed in the service of art compensate for age-related losses on the job, at home, and elsewhere. Working in the art studio, elders learn to take risks and practice ways of reducing self-consciousness; attending concerts and going to museums gives seniors some familiarity with public transportation. Cognitive and personal skills practiced with the arts also broaden decision-making abilities by fostering the attitude that more than one answer to a question or solution to a problem can be right. These skills transfer to other areas of endeavor and settings that make similar physical, mental, personal, and social demands (Haefele, 1962; Nystrom, 1979). Thus, increased self-confidence as a result of artistic competency motivates the elderly to engage in different kinds of complex activities. Competency in the arts can also be applied to larger purposes. "The productive activities of older people... show that the elderly have the potential to contribute to society" (O'Reilly & Caro, 1994, p. 39).

The elderly's involvement in the arts therefore affects physical, sensory, and psychological abilities, fosters self-improvement and personal growth, and transfers to other areas of everyday life. It may also, according to Levy and Langer (1999), increase creative functioning outside of the arts (see especially, pp. 51–52). Training programs that use art to enhance creativity among the elderly have some promise (Parnes, 1999), but such claims, in general, are controversial: the effects are small, and depend on the specific type of training (Rose & Lin, 1984). Training may also depend on whether creativity is defined as "large C," the kind that results in masterpieces of art, or as "little c," as in creatively adapting to life's problems (Ripple, 1999; see also Runco & Dow, 1999; Mayer, 1999). The facilitation of late-life creativity through exposure to art has not, as far as I can determine, been tested. But scientists might be prompted to do so by conducting more research with art and aging artists.

Studies of the elderly's competencies, using materials taken from the arts and testing old artists' abilities with traditional non-art stimuli, along with governmental and other agencies' promotion of art for the elderly, may change the public's assumptions and prejudices about aging, and ultimately society's perception and evaluation of old age. When stereotypes about the old come under increased scrutiny, expectations about aging will become more balanced, fairer, and less pessimistic, lamentations about growing old will be seen as wrong, exaggerated, and only part of the story, and the negative aspects of old age will be de-emphasized. Once younger generations have a greater awareness of the positive links between art and old age, they will become more interested in artistic activities. (For beliefs about old age held by different age groups, see Heckhausen, Dixon, & Baltes, 1989; for changes in long-term views, see Chapter 1.)

But in order for the public to become more engaged with the arts, an early start is essential if these kinds of activities are to "take hold" and be sustained into old age. If art-going and -making are absent in youth, it is difficult to begin them in the middle and later years. "Interests and values remain fairly consistent throughout life .... For an older person to become deeply involved in an arts experience, there usually has to be some long-standing or latent interest" (Susan, 1984, p. 212; see also Kelly, Steinkemp, & Kelly, 1986). The importance of early exposure to the arts was emphasized in the follow-up studies of Terman's gifted children as they aged (Holahan, Sears, & Cronbach, 1995; quotes that follow are from p. 175). "There is remarkable stability in the kinds of activities pursued across the life cycle, especially from middle age onward. Individuals are thus confronted with the challenge of developing, at least by middle age, a set of interests and competencies that they can continue into their later years." However, Holahan et al. pointed out that American society generally holds a less than positive attitude toward leisure, making it difficult for aging citizens to become involved with the arts. "In our contemporary work-oriented culture, leisure is often disparaged as simply nonwork [that] requires little intellectual or physical effort [leading to] ambivalence towards leisure."

Recreational and leisure programs in the arts, along with encouragement, opportunities, and resources, are needed at every age, but particularly among the old whose options are highly limited (O'Reilly & Caro, 1994). But beliefs about aging's deficiencies may dampen, interfere with, hinder, or restrict senior arts programs. On the basis of the research reported here (Part V), there is little reason to think that the elderly's presumed limitations will work against their engagement with the arts.

Whether senior arts programs will continue and expand remains to be seen, though. The influence of the arts on the health and well-being of seniors has not been empirically demonstrated. Does art, as claimed, aid problem-solving and other mental abilities? Is the public's perception of old age changed as a result of visiting exhibitions by old artists? Will a greater awareness of the late-life accomplishments of aging artists result in a more positive attitude toward the aged and aging by the young? Does an exhibition by old artists change young viewers' expectations about old age? Even if these positive effects for art were found, how would they compare with other leisure activities? And would they be on a par with exercise, diet, nutrition, continuing education, and travel? In short, a great deal of scientific research needs to be done on the relationship between the arts and aging.

PROBLEMS AND SOLUTIONS

Accounts of the merits of arts programs for the elderly, outlined above, are not matched by research on their presumed merits. The gap between promises and data was confirmed by a search through several databases. Under the headings of "leisure, old age, and art" covering a 20-year period since the 1970s, there were nearly 100 references to arts programs for the elderly. These covered the advantages of dance, theater, and the like, including arts programs in general, practical advice on administering arts programs, and interviews of old artists. Cited, too, were examples of arts programs in counties, states, and cities, along with references to conferences, directories, resource books, sourcebooks, manuals, reviews, advice, instructions, newsletters, and catalogs published about, for, or by older artists. Numerous essays were written on the relationship between creativity and aging in relation to stress and cultural policy, to arts programs and education, and to geriatric art therapy. Empirical studies, though, were notable by their absence.

An additional 21 publications were identified in another database with the key phrase "artists and age." Seven were about specific old artists, eight referred to discussions of both general and specific arts programs in music, clay, and the visual arts; four cited programs in the arts for seniors in various states and cities; and two noted directories of older artists, resources, and projects. Again, signs of scientific activity were not found. In another database, under "aged and retirement," 79 references to art were cited. Twenty-one discussed general programs in the arts and creativity (27% of the total). Others were scattered over particular areas of the arts, or related art to sports, travel, and second careers. None appeared to have a research base.

Not much more successful was a literature search on style recognition and art (Chapter 15) covering 20 years in *Psychological Abstracts*, the primary database for psychological research. Nearly all citations were to children's reactions to styles of art or their ability to learn and recognize artistic styles; only a few compared young and old viewers of art. In a search of art in applied settings, such as correctional institutions, nursing homes, and hospitals, no empirical studies were found on the effectiveness of arts programming for older residents (Lindauer & Ribner, 1976). It will therefore be a while before research on the relationship between aging and art, little as it is now, will have an impact on arts programs for the elderly, youthful attitudes toward aging, and the elderly's place in society.

In the meantime, numerous suggestions for research were proposed in previous chapters; these are summarized next. The creative accomplishments of aging artists other than painters have yet to be affirmed. Does the creativity of sculptors peak late and continue throughout life, as painters do? What about elderly architects? Aging composers? Similarly, the sanguine self-reports of 60-, 70-, and 80-year-old painters need to be corroborated by aging novelists, dancers, and filmmakers. Future research also has to validate the old-age style for other kinds of visual artists, like architects and sculptors, as well as for non-visual artists, such as writers and composers. Gender is another avenue for investigation. The research on art and creativity was largely based on men (but see Chapter 8). With an increasing number of contemporary long-lived women artists establishing their reputations, and the eventual publication, for some, of complete works (Catalog Raisonnes), their productivity over time can be tallied and compared to men. Age differences in the response to art, too, needs to be investigated with populations other than youthful college undergraduates and 60-year-olds (Chapter 15) so as to include those in their 40s and 80s (Chapter 16). The desirability of longitudinal research has also been stressed. Research can also be improved by showing large reproductions in color, thereby insuring that details, not easily visible in slides and small black and white reproductions, are clearly seen by elderly viewers. In addition, a greater number of examples of art should be shown, rather than pairs or a handful, when comparing young and old art on subject matter, style, and canvas size. Multiple examples take into account the uniqueness of art and artists. Another vital but open question: Whatever holds for doing art, does it also transfer to other complicated tasks that depend on similar skills?

More difficult to investigate are several underlying assumptions about aging, creativity, and the arts. Is the creativity of aging painters and authors the same or like other elderly artists? If so, to what extent does late-life artistic creativity overlap with the creativity of old scholars, scientists, and other professionals outside of the arts? A related assumption that needs to be tested is whether ordinary people in everyday occupations share something of the creativity of artists and other professionals (John-Steiner, 1995). Non-elitist and egalitarian presumptions are admirable but are they correct? This book assumes that creativity is normally distributed throughout the population, and that it works the same way for everyone, albeit at different levels

(Chapter 2). Artists are surely not the only ones who are fiercely dedicated to their work, driven by a sense of what is right, committed to doing their craft well, and energized by distant goals. Aging non-artists also strive to excel in their particular craft, occupation, and job, and struggle mentally and personally to maintain their dedication and worth.

The psychological roots of creativity are another critical premise for empirical assessment. Particularly critical is the role assigned to cognition, which includes a number of relevant traits particularly relevant to the arts, such as perceptual dexterity, imaginative flexibility, richness of imagery, novelty of ideas, and intensity of concentration. But what is the priority of the peculiarly "elderly sparks" of personality and emotion? Late-life creativity, furthermore, is not solely an individual matter of psychological determination; artistic works are not produced in a social-cultural vacuum (Simonton, 1981). Wars, poverty, revolutions, and other traumatic social events determine if, when, and how artists work to and into old age. Large societal forces, as well as interpersonal relations between spouse, friends, other artists, and patrons affect late-life creative expression, and these may be particularly crucial as strength and ambition flag with aging. (For a cross-cultural perspective on creativity, see Raina, 1999, although the focus is neither art nor aging.)

The reader will think of other problematic matters that have not been adequately investigated or have not received sufficient attention in these pages. Progress on bringing aging, creativity, and the arts together is difficult, but given enough time, resolve, and resources, progress can be made. Additional hope also comes from interdisciplinary efforts.

THE CASE FOR INTERDISCIPLINARITY

Knowledge about aging and creativity was furthered by interdisciplinary exchanges between science (mainly psychology and gerontology), the arts (primarily paintings), and the humanities (mostly art history). These connections occurred in many ways. Art historians aided scientific inquiry by nominating aging artists whose productive old ages could be tracked; identified masterpieces whose ages-of-production were traced; specified artists with an old-age style in order to determine a consensus; and provided a lexicon of descriptors of the old-age style in the literature that was evaluated by professionals too (Part IV). Scientific study contributed to humanistic inquiries by estimating the number of artists with an old-age style *and* the number who did not, and the characteristics unique to that special kind of art *and* not to works whose styles did not change as artists grew old. Sixty- to eighty-year-old artists furthered empirical aims by completing a questionnaire on four aspects of creativity, the effects of growing old, and differences between the art of youth and old age. Their reports were quantitatively summarized with statistical formula. Authorities in the humanities, experts in art, and scientific researchers therefore worked together as partners with mutually beneficial outcomes.

Art historians may find it profitable to ponder the significance of several empirical findings about Western art (Part III): the greatest paintings number less than 500; and about 120 artists produced them, of which a handful of 31 were of the highest rank. Scholarly inquiry might be piqued by the data on women artists, who outproduced men during the early but not later years of their careers (Chapter 8). This unexpected finding may prompt a scholarly consideration of gender differences in other kinds of artists in the early stages of their working lives. Inviting comment from experts, too, is why canvas size, but not subject matter, differentiated late-life art so well (Chapter 14). A surge of scholarship should also follow from the empirical identification of painters with an old-age style (Chapters 13 and 14) and the availability of standardized descriptions of their work (Chapters 15 and 16). The lists direct art historians to particular artists and specific works which, through in-depth case studies, can confirm, modify, correct, or expand empirical research on painters with an old-age style and the ways in which this change is revealed. Scholars are also in a better position than they were before now that they have reliable descriptors of the old-age style to identify late-life changes among contemporary painters, and to search for the old-age style among writers, composers, and other artists. Comparisons between the various kinds of art will stimulate scholars to probe similarities and differences between late- life works in diverse artistic modalities.

Scholarly inquiries thus begin where empirically established facts end. Narrative studies can begin with Cézanne and Goya, the only artists who were identified as having an old-age style by a poll of experts and a survey of several major published sources. More ambi- tious art historians might concentrate on the 38 artists selected by a panel of experts as having an old-age style (Cézanne, Constable, Degas, and others), and compare them with the 68 who did not (Albers, Bingham, Cole, Giotto, Gorky, and the rest; Chapters 13 and 14). Or they can take a second look at artists who have received mixed evalu- ations in the literature (Corot, Degas, El Greco, and so on). Scholarship might be directed at these artists' works, biographies, or their times. For example, Eakins and Tiepolo, artists with and without an old-age style, respectively, could be compared on their working habits, professional associates, family circumstances, economic conditions, and the like. These and other scholarly inquiries—for example: the waxing of the old-age style in the past and its apparent waning in the present; why some artists, like Goya, Michelangelo, and Rembrandt, continued to be creative late in life and others did not; and how the old-age art of modern and post-modern artists differ from the past—would not have come to scholars' attention without quantitative research leading the way.

Aging artists, too, provided materials for studies as well as participated in the research. They contributed concrete, dramatic, and compelling examples of late-life art that gave substance to scientific generalizations and abstract humanistic discourses. They also offered personal testimony, so-called anecdotal data, on why their late-life work was affected by aging, and if not, why not. Artists and art are also a source of hypotheses about aging and these can be tested. Consider their depictions of the stages of man (Chapter 2), which are essentially snapshots of human development. They portray the various ways in which aging has been conceptualized, and whether these have been constant or variable over time. The recognition of the value of these pictorial representations of aging and old age, a thorough analysis by art historians, and tying them to current scientific thinking about aging, remain to be done.

Both artists and art historians serve other empirical purposes. They act as watchdogs over scientific efforts, evaluating the way artistic materials are used in research, pointing out inappropriate selections, distortions in their meaning, and violations of the artists' intent. They also insure that statistical aggregates of aging artists and late works do not lose sight of unique artists, extraordinary works, and exceptions. Scholars and artists are sensitive to the danger of scientific investigations treating dissimilar old artists and diverse sets of late works as if they were all from the same mold. By heeding scholars' and artists' insistence on not losing the individual qualities of late works and specific features of particular artists, scientists can avoid excessive quantification and brutal overgeneralizations. By warning scientists about superficial interpretations and glib conclusions, scholars and artists also serve their own needs, insuring that their fields not become irrelevant or subordinate to scientific inquiry.

Discussions by scholars and examples from artists fuel scientific studies. Empirical efforts, in turn, further scholarly debate and affect artists' thinking about and ways of expressing concepts. Critical essays and new forms of art then become a source of additional leads and hypotheses for research. These prompt further critical discussions from humanists and counter-examples from artists, and so on, in a fruitful cycle of circularity. Thus, a hypothesis-driven, empirically oriented, and quantitative approach to old artists, late works, the old-age style, and elderly arts audiences suggest scholarly inquiries and personal comments from artists, either directly or indirectly in their work. Do the biographies of and reports from aging artists complain about the impact of old age on their art? Do the personal records and narrative accounts of artists with physical and other handicaps reveal if and how they worked around them? Were aging artists forced to change the way they worked, the materials they used, the medium they worked in, or the techniques they used? Recall that reports from contemporary artists refuted the deleterious affects of physical and sensory losses on their art. The quantitative framework generated by scientists was therefore a backdrop against which individual works and artists' words were more closely examined. Did they fit the general pattern, and if not, why did they not, and what do the exceptions say about the overall results?

Anecdotal reports by artists are incomplete. So too are scholarly discourses on individual artists and specific works. Likewise, isolated facts and data-bound scientific conclusions restricted to statistical formulations, graphs, and tables are insufficient. Taken together, though, disparate, competing, and incompatible views from artists, humanists, and scientists can arrive at a more complete picture. Interdisciplinary research is as complicated as laboratory studies but the rewards are great. An interdisciplinary approach prevents scientists, scholars, and artists from "going their own way," overlooking each other's contributions and thereby missing an opportunity to learn more about late-life creativity and old age. Regrettably, interdisciplinarity is hindered by miscommunications, reservations, barriers, and territorial imperatives between fields of inquiry (Chapter 2). In fairness, it is not easy to integrate the methodology of science with the qualitative mode of inquiry in the humanities and the expressive forms of communication in the arts. The synthesis is made more difficult by the clamor of other disciplines vying to be heard, including psychology in both its empirical and clinical versions, gerontology, art history, and literary criticism, not to speak of competing voices from biology, the health care professions, art therapy, art education, and the fields of recreation and leisure, along with the pragmatic demands of senior centers and the programmatic concerns of government agencies concerned with arts for the aged.

An exchange of views, examples, arguments, and information between disciplines is useful, but it also breeds contentious issues. Intra-disciplinary connections, too, are problematic. Within empirical psychology alone, late-life development requires a synthesis of sensory, perceptual, and cognitive fields of inquiry, as well as input from the study of personality, motivation, emotions, and attitudes. Different lines of evidence, argument, and analyses espoused by various discipline and their sub-disciplines invariably lead to conflict. Which route to knowledge is best? Revelatory examples, compelling anecdotes, personal testimony, or computer-generated statistics? Different means of validation are not easily reconciled.

The forging of interdisciplinary ties has long been proposed by historians, philosophers, social critics, intellectuals, and other commentators. Cycles of argument and counter-argument, usually tendered without evidence or alluding to research selectively, have not moved the debate along. Interdisciplinary possibilities have not been concretely demonstrated by studies.

This book did so by meeting two essential criteria for interdisciplinarity. The first was to join disparate disciplines by focusing on topics with a common nexus, namely, aging, creativity, and art (Part I). The second step was to demonstrate the usefulness of this convergence by conducting empirical studies that combined humanistic scholarship in the arts, on the one hand, with artistic examples, research materials, and testimonies, on the other hand, along with the participation of scholars and artists in research (Parts II–IV). The seemingly endless and contentious debates that have dogged the science-humanities schism for generations (Chapter 2) was reduced by conducting scientific research that was cognizant of discussions in the humanities and accomplishments in the arts (Lindauer, 1995, 1998b, see especially the introduction, and *Science, Art, and the Humanities*, n.d.; see also Lindauer, 1978).

The lesson is clear. When scientists study old age and become more knowledgeable about old artists and late art, and gain a greater awareness of the multiple paths of late-life achievement, it is difficult to accept the decline position on creative development unequivocally or take an overly pessimistic view of aging (Chapter 3). Similarly, when art historians and artists become cognizant of the scientific evidence for decline and stagnation among the aged, they will be less sanguine about the course of late-life artistic creativity. Scientific, scholarly, and artistic avenues of inquiry therefore proceed best in tandem rather than on parallel (if not opposing) tracks.

Interdisciplinary efforts can proceed with some confidence, given their successful forays in this book. Controlled procedures, statistical analyses, quantitative instruments, and group averages were sensitively joined with discursive essays, qualitative judgments, subjective opinions, and personal comments by humanists and artists. The two also served as research participants, and their responses were not so idiosyncratic or individualistic as to be empirically useless or irrelevant. The judgments of art experts and the personal testimonies of artists were obtained systematically by relying on pre-tested and structured empirical procedures. Works by artists with an old-age style were compared by laypersons in a quasi-experimental design to a "control group," that is, to works without this change. The quantitative findings on the relatively minor affects of physical and sensory losses on artists' late works were buttressed by the artists' own words. Group averages described the peak years of creative productivity in general (Chapter 7) while profiles of individual artists illustrated the larger pattern—and exceptions (Chapter 8). Once the performance of artists as a whole was established statistically, particular artists were juxtaposed against the larger framework. Similarly, quantitative generalizations about the late age at which artistic creativity reached its zenith, continued, leveled off, and declined were supplemented by examples of specific artists who either did or did not reflect this general pattern. Along similar lines, a qualitative instrument, an open-ended questionnaire, was completed by old living artists and systematically tabulated by means of a formal content analysis. The credibility of this statistical weeding instrument was enhanced by extensive quotes from artists (Chapters 9 and 10); numerical averages and personal reports were intertwined. Other examples of interdisciplinary cooperation included experts' identification of historical artists and works that exemplified the old-age style, and following up the implications of these findings in questionnaires completed by living old artists (Chapter 13). Exchanges across disciplines were also illustrated by the study of the ages at which masterpieces were produced, which drew upon Catalogue Raisonnes and catalogs of art exhibitions compiled by experts (Chapters 7 and 8).

A LAST WORD: WHEREFORE ART?

"What is observed to be normal [in old age] does not necessarily tell us what is possible," asserted Baltes and Baltes (quoted in

Schulz & Heckhausen, 1996, p. 702). By "normal" the Baltes were referring to the predominance of unhealthy and institutionalized individuals in research on old age and the absence of more competent elders. Baltes and Baltes did not deny the reality of losses with age. "In old age, declines in both physical and cognitive functioning are evident." Nevertheless, Baltes and Baltes were optimistic. "The rate of decline and domains in which decline occurs is quite variable and may not be irreversible." Their optimism, which they directed to healthy and non-institutionalized old adults, also applies to aging creative artists, elderly participants in the arts, and aging arts audiences.

Exemplary performance in old age is certainly not limited to aging artists and artistic contexts. Outstanding accomplishments are achieved by many people and in various fields outside of the arts (Chapter 1). Successful aging is not fulfilled by engaging only in the arts (Chapter 2). Art is not the only endeavor that requires originality and innovation. Productive aging is not just about old art, old artists, and old art audiences; creativity is not solely about art; and art is not simply a psychological or cognitive matter. Although this book is about art, its implications are much broader. The abilities of the old, whether expressed by art or otherwise, do not necessarily stagnate, fall into disrepair, or disappear; skills persist, improve, emerge anew, become energized, and take different forms in old age.

Accomplishments in art, though, provide compelling evidence for the strengths of aging and lend empirical support to a positive view of late-life potential (Uttal & Perlmutter, 1989). Aging creative artists and older arts audiences affirmed their ability to adapt to and meet the demands of old age. The rest of us can therefore hope, with some justification, that creative, artistic, psychological, physical, and sensory competencies will continue into the later years. Old age can be a period of stagnation and decline, but also a time of stability as well as growth. Creativity in old age, like wisdom, can be absent or rare, or it may flower.

Research on the relationship between art, aging, and late-life creativity converged on one major point: Growing old is not necessarily bad. Optimism is justified. The dire predictions of the decline model of aging and creativity (Chapter 3) have been assuaged if not reversed: old age can no longer be unequivocally associated with loss; aging can also be a time of gains. Yes, additional debate, other examples, and more facts are needed; the weight of opposing evidence is great. But this book signals a promising start to revising and updating centuries of a perversely negative interpretation of the course of creativity and aging in general, and late-life artistic involvement and creativity in particular.

References

Abra, J. (1989). Changes in creativity with age: Data, explanations, and further predictions. *International Journal of Aging and Human Development, 28*, 105–125.

Abra, J. (1995). Do the muses dwell in Elysium? Death as a motive for creativity. *Creativity Research Journal, 8*, 205–217.

Achenbaum, W. A. (1989). Foreword: Literature's value in gerontological research. In P. V. D. Bagnell & P. S. Sapir (Eds.), *Perception of aging in literature: A cross-cultural perspective* (pp. viii–xxv). NY: Greenwood.

Achenbaum, W. A., & Kusnerz, P. A. (1982). *Images of old age in America: 1790 to the present*. Institute of Gerontology, The University of Michigan-Wayne State University. (Originally published 1978.)

Adams-Price, C. (1998a). Aging, writing, and creativity. In C. Adams-Price (Ed.), *Creativity and aging: Theoretical and empirical perspectives* (pp. 269–287). NY: Springer.

Adams-Price, C. (Ed.) (1998b). *Creativity and aging: Theoretical and empirical perspectives*. NY: Springer.

Ager, C. L., White, L. W., Mayberry, W. L., Crist, P. A., & Conrad, M. E. (1981–82). Creative aging. *International Journal of Aging and Human Development, 14*, 67–76.

Ahmad, P. J. (1985). Visual art preference studies: A review of contradictions. *Visual Arts Research, 11*, 100–107.

Albert, A. (1975). Toward a behavioral definition of genius. *American Psychologist, 30*, 140–151.

Allen Memorial Art Museum. (1962). Youthful works by great artists: A symposium. *Allen Memorial Art Museum Bulletin Oberlin College, 20*, 90–168.

Alpaugh, P. K., & Birren, J. E. (1975). Are there sex differences in creativity across the adult life span? *Human Development, 18*, 461–465.

Alpaugh, P. K., & Birren, J. E. (1977). Variables affecting creative contributions across the adult life span. *Human Development, 20*, 240–248.

Alpaugh, P. K., Parham, I. A., Cole, K. D., & Birren, J. E. (1983). Creativity in adulthood and old age. *Educational Gerontology, 1*, 17–40.

Andersson, E., Berg, S., Lawenius, M., & Ruth. (1989). Creativity in old age: A longitudinal study. *Aging, 1*, 159–164.

Arcamore, A., & Lindauer, M. S. (1974). Concept learning and the identification of poetic style. *Psychological Reports, 35*, 207–210.

Arenberg, D. (1978). Differences and changes with age on the Benton Visual Test. *Journal of Gerontology, 33*, 534–540.

Arlin, P. K. (1975). Cognitive development in adulthood. A fifth stage? *Developmental Psychology, 11*, 602–606.

Arnason, H. H., with Wheeler, D. (1986). *History of modern art* (3rd ed.). NY & Englewood Cliffs, NJ: Prentice-Hall & Harry N. Abrams.

Armour, R. A. (1980). *Film: A reference guide*. Westport, CN: Greenwood Press.

Arnheim, R. (1986a). On the late style. In R. Arnheim (Ed.), *New essays on the psychology of art* (pp. 285–293). Berkeley, CA: University of California Press. (Originally published in 1978.)

Arnheim, R. (1986b). The other Gustav Theodor Fechner. In R. Arnheim (Ed.), *New essays on the psychology of art* (pp. 39–49). Berkeley, CA: University of California Press.

Arnheim, R. (1990). On the late style. In M. Perlmutter (Ed.), *Late life potential* (pp. 113–120). Washington, DC: Gerontological Society of America.

(1990). A self-taught artist who began drawing at 85 [Bill Trayler]. *The Chronicle of Higher Education, 36* (June 13), *B48.*

Auden, W. H. (1976). Old people's home. In E. Mendelson (Ed.), *W. H. Auden collected poems.* London: Faber & Faber.

Azar, B. (1998). Older workers need not be left behind by technology. *American Psychological Association Monitor, 29,* 23–24.

Bagnell, P. v. D., & Soper, P. S. (Eds.). (1989). *Perception of aging in literature: A cross-cultural perspective* (pp. xxiii–xxv). NY: Greenwood.

Baer, J. (1999a). Domains of creativity. In M. A. Runco & S. R. Spitzker (Eds.), *Encyclopedia of creativity* (Vol. 1, pp. 591–596). San Diego: Academic Press.

Baer, J. (1999b). Gender differences. In M. A. Runco & S. R. Spitzker (Eds.), *Encyclopedia of creativity* (Vol. 1, pp. 753–758). San Diego: Academic Press.

Balin, S. (1994). *Achieving extraordinary ends: An essay on creativity.* Norwood, NJ: Ablex.

Balkema, J. B. (1986). *The creative spirit: An annotated bibliography on the arts, humanities & aging.* Washington, DC: National Council on the Aging.

Baltes, P. B., Smith, J., Staudinger, U. M., & Sowarka, D. (1990). Wisdom: One facet of successful aging? In M. Perlmutter (Ed.), *Late life potential* (pp. 63–81). Washington, DC: The Gerontological Society of America.

Baltes, P. B., & Staudinger, U. M. (1995). Wisdom. In G. L. Maddox (Ed.), *Encyclopedia of aging* (pp. 971–974). NY: Springer.

Baltes, Paul, B., & Staudinger. (2000). Wisdom: A metaheuristic (pragmatic) to orchestrate mind and virtue toward excellence. In M. E. P. Seligman & M. Csikszentmihalyi (Eds.), Positive psychology: An introduction. Special Issue on happiness, excellence, and optimal human functioning. *American Psychologist, 55,* 122–136.

Beard, G. M. (1874). *Legal responsibility in old age: Based on research into relations of age to work.* NY: Russell's American Steam Printing House.

de Beavoir, Simone. (1950). *The coming of age* (P. O'Brian, Trans). NY: G. D. Putnam.

Becker, G. (2001). The association of creativity and psychopathology: Its cultural-historical origins. *Creativity Research Journal, 13,* 45–53.

Behrens, J. (1990). Pirxouttes, paint brushes, and prescriptions. *The Journal of NIH Research, 2,* 46–47.

Benn, G. (1955). Artists and old-age. *Partisan Review, 22,* 297–319.

Berg, G., & Gadow, S. (1978). Toward more human meanings of aging: Ideals and images from philosophy and art. In S. F. Spicker, R. M. Woodward, & D. D. van Tassell (Eds.), *Aging and the elderly: Humanistic perspectives in gerontology* (pp. 83–92). Atlantic Highlands, NJ: Humanities Press.

Bergmann, K. (1972). *The aged: Their understanding and care.* London: Wolfe Publications.

Berlind, R. (Ed.). (1994). Art and old age. *Art Journal, 53,* 7–85.

Berlyne, D. E. (1974). *The new experimental aesthetics.* NY: Hemisphere.

Berman, A. (1983). When artists grow old. *Art News, 82,* 76–83.

Berman, H. J. (1994). *Interpreting the aging self: Personal journals of later life.* NY: Springer.

Bertman, S. L. (1989). Aging grace: Treatments of the aged in the arts. *Death Studies, 13,* 517–535.

Bink, M. L., & Marsh, R. L. (2000). Cognitive regularities in creative activity. *Review of General Psychology, 4,* 59–78.

Binstock, R. H., & George, L. K. (1985). *Handbook of aging and the social sciences* (2nd ed.). San Diego: Academic Press.

Binstock, R. H., & George, L. K. (1990). *Handbook of aging and the social sciences* (3rd ed.). San Diego: Academic Press.

Binstock, R. H., & George, L. K. (1996). *Handbook of aging and the social sciences* (4th ed.). San Diego: Academic Press.

Birren, J. E. (1985). Age, competence, creativity, and wisdom. In N. N. Butler & H. P. Gleason (Eds.), *Productive aging* (pp. 29–34). NY: Springer.

Birren, James, E. (1990). Creativity, productivity, and potentials of the senior scholar. Exploring the feasibility of an institute of senior scholars [Special issue]. *Gerontology and Geriatrics Education, 11,* 27–44.

Birren, J. E., & Deutchman D. E. (Eds.). (1991). Introduction: Exploring the feasibility of an Institute of Senior Scholars. *Gerontology and Geriatric Education, 11,* 1–159.

Birren, J. E., & Schaie, W. W. (1977). *Handbook of the psychology of aging.* NY: Van Nostrand Reinhold.

Birren, J. E., & Schaie, K. W. (1985). *Handbook of the psychology of aging* (2nd ed.). NY: Van Nostrand Reinhold.

Birren, J. E., & Schaie, K. W. (1990). *Handbook of the psychology of aging* (3rd ed.). San Diego: Academic Press.

Birren, J. E., & Schroots, J. J. F. (Eds.). (2000). *A history of geropsychology in autobiography.* Washington, DC: American Psychological Association.

Blythe, R. (1979). *The view in Winter: Reflections on old age.* NY: Penguin.

Boden, M. A. (Ed.). (1994). *Dimensions of creativity.* Cambridge, MA: MIT.

Boggs, J. S. (1977–78). Edgar Degas in old age. In W. Hood (Ed.), *Artists and old age: A symposium,* Allen Memorial Art Museum Bulletin Oberlin College, *35,* 57–67.

Booth, W. (1992). *The art of growing older: Writers on living and aging.* Chicago: University of Chicago Press.

Bornstein, M. H. (1984). Developmental psychology and the problem of artistic change. *Journal of Aesthetics and Art Criticism, 43,* 131–145.

Bowerman, W. G. (1947). *Studies in genius.* NY: Philosophical Library.

Bramwell. (1947). Galton's "Hereditary Genius" and the following generations since 1869. *Eugenics Review, 39,* 146–153.

Breskin, M. (1954). Foreword. *Man and his years: Catalogue of an exhibition by the Baltimore Museum of Art.* Baltimore, MD.

Brinkmann, A. E. (1925). *Spatwerke grosser Meister.* [Late works of great masters.] Frankfurt am Main. (Peter Ryan, Trans.). University of Michigan Seminar by Dr. Marvin Eisenberg on "The Artist in Old Age."

Brinkmann, A. E. (1984). Earliest and latest works of great artists. *Gazette des Beaux-Arts, 46,* 273–284.

Bromley, D. B. (1956). Some experimental tests of the new effect of age on creative intellectual output. *Journal of Gerontology, 11,* 74–82.

Bryson, N. (1983). *Vision and painting: The logic of the gaze.* London: MacMillan.

Bullough, V., Bullough, B., & Maddalena, M. (1978). Age and achievement: A dissenting view. *The Gerontologist, 18,* 584–589.

Bullough, V. L., Voght, M., Bullough, B., & Kluckhohn, L. (1980). Longevity and achievement in Eighteenth Century Scotland. *Omega, 1,* 115–119.

Butler, R. N. (1963). Life review: The interpretation of reminiscence in the elderly. *Psychiatry: Journal for the Study of Interpersonal Processes, 26,* 7.

Butler, R. N. (1973). The creative life and old age. In E. Pfeiffer (Ed.), *Successful aging: A conference report* (pp. 97–107). Durham, NC: Duke University Press.

Butler, R. N. (1974). Successful aging and the role of the life review. *Journal of the American Geriatrics Society, 22,* 529–535.

Butler, R. N. (1990) The contribution of late-life creativity to society. In J. E. Birren & D. E. Deutchman (Eds.), Exploring the feasibility of an Institute of Seniors (pp. 45–51). *Gerontology and Geriatric Education, 11,* 1–159.

Butler, R. N., & Gleason, H. P. (Eds.). (1985). *Productive aging: Enhancing vitality in later life.* NY: Springer.

Byers, M. (1982, January). Older artists create an "ageless" beauty. *Smithsonian News Service,* Smithsonian Institution (pp. 1–5).

Cahill, P. (1981). *The arts and older Americans. A catalog of program profiles.* Washington, DC: National Council on the Aging.

Cattell, J. Mc. (1903). A statistical study of eminent men. *Popular Science Monthly, 62,* 359–377.

Champlin, J. B., Jr. (1914). *Cyclopedia of painters and paintings,* (4 vols). NY: Charles Scribner's Sons.

Charles, D. C. (1977). Literary old age: A browse through history. *Educational Gerontology, 2,* 237–253.

Charness, N. (1989). Age and expertise: Responding to Talland's challenge. In L. W. Poon, D. C. Rubin & B. A. Wilson (Eds.), *Everyday cognition in adulthood and later life* (pp. 437–456). Cambridge: Cambridge University Press.

Chudacoff, H. P. (1992). *How old are you? Age consciousness in American culture.* NJ: Princeton University Press.

Clark, E. L. (1916). *American men of letters.* NY: Columbia University Press.

Clark, K. (1972). *The artist grows old (The Rede Lectures, 1970).* New York: Cambridge University Press.

Clayton, V. (1975). Erikson's theory as it applies to the aged. Wisdom as contradictive cognition. *Human Development, 2,* 119–128.

Clayton, V. (1987). Wisdom. In G. L. Maddox (Ed.), *The encyclopedia of aging* (pp. 696–710). NY: Springer.

Clayton, V. P., & Birren, J. E. (1980). The development of wisdom across the life span: A reexamination of an ancient topic. *Life Span Development and Behavior, 1,* 103–135.

Clayton. V. P., & Overton, W. F. (1970). Concrete and formal thought process in aged. *International Journal of Aging and Human Behavior, 7,* 232–245.

Coffman, G. R. (1934). Old age from Horace to Chaucer: Some literary affinities and adventures of an idea. *Speculum, 9,* 249–277.

Cohen, G. D. (1997). Contemplating creativity. *AARP Bulletin, 36,* 2, 5.

Cohen, S. (1986). Artistic creativity across the adult life-span: An alternative approach. *Interchange, 17,* 1–16.

Cohen-Shalev, A. (1989a). *The aging psyche and the question of artistic change: An examination of their relationship in one career [on Verdi].* Unpublished manuscript.

Cohen-Shalev, A. (1989b). Old age style: Developmental changes in creative production from a life-span perspective. *Journal of Aging Studies, 3,* 21–37.

Cohen-Shalev. (1992). Self and style: The development of artistic expression from youth through midlife to old age in the works of Henrik Ibsen. *Journal of Aging Studies, 6,* 289–299.

Cohen-Shalev, A. (1993). Mondrian and the "boogie-woogie": Interruption of inner developmental logic or completion in old age? *International Journal of Aging and Human Development, 38,* 1–8.

Cohen-Shalev, A., & Rapoport, T. (1990). *"Completion of being" and "fragmentation of inner reality": Two views of aging in Ingmar Bergman's "Wild Strawberries" and "Franny and Alexander."* Unpublished manuscript.

Cohen, E. S., & Kruschwitz, A. L. (1990). Old age in America represented in nineteenth and twentieth century popular sheet music. *Gerontologist, 30,* 345–354.

Cole, S. (1979). Age and scientific performance. *American Journal of Sociology, 84,* 958–977.

Cole, T. R. (1992a). The humanities and aging: An overview. In T. R. Cole, D. D. Van Tassel, & R. Kastenbaum (Eds.). *Handbook of the humanities and aging* (pp. x–xviv). NY: Springer.

Cole, T. R. (1992b). *Journey of life: A cultural history of aging in America.* Cambridge, England: Cambridge University Press.

Cole, T. R., & Winkler, M. C. (1994). *The Oxford book of aging: Reflections of the journey of life.* NY: Oxford University Press.

Coleman, P. G. (1992). Personal adjustment in late life: Successful aging. *Reviews in Clinical Gerontology, 2,* 67–78.

Commons, M. L., Richards, F. A., & Armon, C. (Eds.). (1984). *Beyond formal operations: Late adolescent and adult cognitive development.* NY: Praeger.

Cooke, N. J. (1990). Practice makes perfect—Or does it? Review of M. T. H. Chi, R. Glaser, & M. J. Farr (Eds.), *The nature of expertise. Contemporary Psychology, 35,* 251–253.

Cornelius, D. K. (1964). *Cultures in conflict: Perspectives on the Snow-Leavis controversy.* Chicago: Scott, Foresman.

Corsa, Helen, B. (1973). Psychoanalytic concepts of creativity and aging: The fate of creativity in mid-years and old age. *Journal of Geriatric Psychiatry, 6,* 169–183.

Covey, H. C. (1989). Old age portrayed by the ages-of-life models from the Middle Ages to the 18th century. *Gerontologist, 29,* 692–698.

Covey, H. C. (1991). *Images of older people in Western art and society.* NY: Praeger.

Cowley, M. (1980). *The view from 80.* NY: Wiley.

Cox, C. (1926). *Genetic studies of genius. vol. 2. The early mental traits of three hundred geniuses.* Stanford, CA: Stanford University Press.

Craik, F. I. M., & Simon, E. (1980). Age differences in memory: The role of attention and depth of processing. In L. W. Poon, J. L. Fozard, L. S. Cermak, D. Arenberg, & L. W. Thompson (Eds.), *New directions in memory and aging* (pp. 95–112). Hillsdale, NJ: Lawrence Erlbaum.

Craik, F. I. M., & Salthouse, T. A. (Eds.). (1992). *The handbook of aging and cognition.* Hillsdale, NJ: Erlbaum.

Cravens, H. (1992). The Terman genetic studies of genius, 1920–1950. *American Psychologist, 47,* 183–189.

Cropley, A. J. (1999). Definition of creativity. In M. A. Runco & S. R. Spitzker (Eds.), *Encyclopedia of creativity* (vol. 1, pp. 511–524). San Diego: Academic Press.

Csikszentmihalyi, M. (1989). Review of D. K. Simonton's Scientific genius: A psychology of science (1988). *Contemporary Psychology, 34,* 981–982.

Csikszentmihalyi, M. (1991). An investment theory of creativity and its development: Commentary. *Human Development, 34,* 32–34.

Csikszentmihalyi, M. (1997). *Finding flow.* NY: Basic Books.

Cultler, S. J., & Hendricks, J. (1990). Gender differences in the arts. In R. H. Binstock & L. K. George (Eds.), *Handbook of aging and the social sciences* (4th ed., pp. 169–185). San Diego: Academic Press.

Cummings, E. (1979). *Growing old: The process of disengagement.* NY: Arno Press. (Originally published in 1961 by Basic Books.)

Dacey, J.(1999). Aging. In M. A. Runco & S. R. Spitzker (Eds.), *Encyclopedia of creativity* (vol. 1, pp. 309–322). San Diego: Academic Press.

DeAngelis, T. (1989). Elder tales stress later-life challenges. *American Psychological Association Monitor, 20*, 40–41.

de Grazia, S. (1961). The uses of time. In R. W. Kleemeier (Ed.), *Aging and leisure: A research perspective into the meaningful use of time* (pp. 113–153). NY: Oxford University Press.

Dennis, W. (1955). Variations in productivity among creative works. *Scientific Monthly, 80*, 277–278.

Dennis, W. (1956). Age and achievement: A critique. *Journal of Gerontology, 11*, 331–333.

Dennis, W. (1966). Creative productivity between the ages of 20 and 80 years. *Journal of Gerontology, 21*, 1–8.

Denny, N. W. (1982). Aging and cognitive change. In B. Wolman (Ed.), *Handbook of developmental psychology* (pp. 807–827). Englewood Cliffs: Prentice-Hall.

Diamond, A. M., Jr. (1984). An economic model of the life-cycle research productivity of scientists. *Scientometrics, 6*, 189–196.

DiGiammarino, M., Hanlon, H., Kassing, G., & Libman, K. (1992). Arts and aging: An annotated bibliography of selected resource materials in art, dance, drama and music. *Activities, Adaptation, and Aging, 17*, 39–51.

DiMaggio, P., & Useem, M. (1980). *Audience studies of the performing arts and museums. A critical review.* Washington, DC: National Endowment for the Arts.

Dohr, J. H., & Forbes, L. A. (1986). Creativity, arts, and profiles of aging. *Educational Gerontology, 12*, 123–138.

Dreher, R. E. (1968). Esthetic responses to paintings: As originals, kodachrome, prints and black-and-white photographs. *Proceedings, 76th Annual Convention, America Psychological Association, 3*, 451–452.

Dudek, S. Z., & Croteau, H. (1998). Aging and creativity in eminent architects. In C. Adams-Price (Ed.), *Creativity and aging: Theoretical and empirical perspectives* (pp. 117–149). NY: Springer.

Dudek, S. Z., & Hall, W. B. (1991). Personality consistency: Eminent architects 25 years later. *Creativity Research Journal, 4*, 213–231.

Dunforth, P. (1989). *A biographical dictionary of women artists in Europe and America since 1850.* Philadelphia, PA: University of Pennsylvania Press.

Edel, L. (1977–78). Portrait of the artist as an old man. *The American Scholar, 47*, 52–62.

Edel, L. (1979). Portrait of the artist as an old man. In D. D. Van Tassel (Ed.), *Aging, death, and the completion of being* (pp. 193–214). Philadelphia, PA: University of Pennsylvania Press.

(1979). *Education: An arts/aging answer.* Washington, DC: The National Council on the Aging.

(1968). *Encyclopedia of world art* (25 vol.). NY: McGraw-Hill.

Erikson, E. (1963). *Childhood and society* (2nd ed.). NY: Norton.

Erikson, E. (1982). *The life cycle completed.* NY: Norton.

Erikson, E., Erikson, J. M., & Kivnik, H. Q. (1986). *Vital involvement in old age.* NY: Norton.

Erikson, J. (1988). *Wisdom of the senses: The way of creativity.* NY: W. W. Norton.

Eysenck, H. J. (1995). *Genius: The natural history of creativity.* Cambridge: Cambridge University Press.

Eyben, E. (1989). Bibliography. Old age in Greco-Roman antiquity and early Christianity: An annotated selective bibliography. In T. M. Falkner & J. de Luce (Eds.), *Old age in Greek and Latin literature* (pp. 230–251). NY: State University of NY Press.

Falkner, T. M., & Luce, de J. (Eds.). (1989). *Old age in Greek and Latin literature*. NY: State University of NY Press.

Fallis, R. C. (1989). Grow old along with me: Images of older people in British and American Literature. In P. v. D. Bagnell & P. S. Sopir (Eds.), *Perception of aging in literature: A cross-cultural perspective* (pp. 35–50). NY: Greenwood.

Famishetti, R. A. (1988). Older adults and the arts: Agenda for knowledge and action.

Feinstein, R. (1988). *Enduring creativity*. NY: Whitney Museum of American Art.

Feldman, F. (1992). *I am still learning: Late works by masters*. Washington, DC: National Gallery of Art.

Fischer, D. H. (1977). *Growing old in America* (Expanded ed.). NY: Oxford University Press.

Fisher, B. J., & Specht, D. K. (1999). Successful aging and creativity in later life. *Journal of Aging Studies, 13*, 457–472.

Fisher, L. E. (1985). *Masterpieces of American painting*. NY: Exeter/Bison.

Foote, J. (1994, August 14). The art of aging gracefully. *The Oregonian*, D1, D4.

Foster, M. T. (1992). Experiencing a "creative high." *Journal of Creative Behavior, 28*, 29–39.

Franklin, M. B. (2001). The artist speaks: Sigmund Koch on aesthetics and creative work. *American Psychologist, 56*, 445–452.

Galton, F. (1952). *Hereditary genius: An inquiry into its laws and consequences*. NY: Horizon Press. (First edition 1869: London: Macmillan)

Gantner, J. (1965). Der Alte Kunstler [The old artist]. From *Festschrift fur Herbert von Einmen* (pp. 71–76). Translated for the University of Michigan seminar, "The artist in old age," Marvin Eisenberg.

Gardner, H. (1973). *The arts and human development*. NY: Wiley.

Gardner, H. (1974). The contribution of color and textures to the detection of painting style. *Studies in Art Education, 15*, 57–62.

Gardner, H. (1983). *Frames of mind*. NY: Basic Books.

Garfield, S. L. (1998). The future and the scientist–practitioner split. *American Psychologist, 53*, 1231–1232.

Gedo, M. M. (1984). The twilight of the Gods. In J. E. Gedo & G. H. Pollock (Eds.), *Psychoanalysis: The vital issues. Vol. 1. Psychoanalysis as an intellectual discipline. Emotion and Behavior Monographs*, No. 2 (pp. 321–360). NY: International Press.

Getzels, J. W. (1990). Creativity. In R. M. Thomas (Ed.), *The encyclopedia of human development and education: Theory, research, and studies* (pp. 291–299). Oxford: Pergamon.

Getzels, J. W., & Csikzentmihalyi, M. (1970). Concern for discovery: An attitudinal component of creative production. *Journal of Personality, 38*, 91–105.

Giambra, L. (1987). Concept learning. In G. L. Maddox (Ed.), *The encyclopedia of aging* (pp. 139–140). NY: Springer-Verlag.

Goff, K. (1993). Creativity and life satisfaction of older adults. *Educational Gerontology, 10*, 241–250.

Goldman, C. (1991). Late bloomers: Growing older or still growing? *Generations, 10*, 41–44.

Goodich, M. W. (1989). *From birth to old age: The human life cycle in medieval thought, 1250–1350*. Lanham, MD: University Press of America.

Graf, P. (1990). Life-span changes in implicit and explicit memory. *Bulletin of the Psychonomic Society, 28*, 353–358.

Greenberg, P. (1987). *Visual arts and older people: Developing quality programs*. Springfield, IL: Charles C. Thomas.

Gruber, H. E. (1974). *Darwin on man: A psychological study of scientific creativity*. NY: Dutton.

Grumbach, D. (2000). What old age is really like. *The New York Times, 148*, 15.

(1998). Guidelines of the evaluation of dementia and age-related cognitive decline. *American Psychologist, 53*, 1298–1303.

Guilford, J. P. (1950). Creativity. *American Psychologist, 5*, 444–454.

Guilford, J. P. (1967). *The nature of human intelligence.* NY: Macmillan.

Gutmann, D. L. (1987). *Reclaimed powers: Toward a new psychology of men and women in later life.* NY: Basic Books.

Haefele, J. W. (1962). *Creativity and innovation.* NY: Reinhold.

Hamilton, G. H. (1984). The dying of the light: The late works of Degas, Monet, and Cézanne. In J. Rewald & F. Weitzenhoffer (Eds.), *Aspects of Monet: A symposium on the artist's life and times* (pp. 220–241). NY: Abrams.

Hall, G. S. (1922). *Senescence: The last half of life.* NY: Reinhold.

Harker, J. O., & Riege, W. H. (1985). Aging and delay effects on recognition of words and designs. *Journal of Gerontology, 40*, 601–604.

Hartt, F. (1985). *A history of art: Painting, sculpture, architecture* (2nd ed., Vol. II). Englewood cliffs, NJ and NY: Prentice-Hall & Harry N. Abrams.

Heckhausen, J., Dixon, R. A., & Baltes, P. B. (1989). Gains and losses in development throughout adulthood as perceived by different adult age groups. *Developmental Psychology, 25*, 109–121.

Held, J. S. (1987). Commentary. In D. Rosand (Ed.), Old-age style. *Art Journal, 46*, 127–133.

Hess, T. M., & Wallsten, S. M. (1987). Adult age differences in the perception and learning of artistic style categories. *Psychology and Aging, 2*, 243–253.

Hiemstra, R. (1982). Elderly interests in the expressive domain. *Educational Gerontologist, 8*, 143–154.

Hocevar, D. (1981). Measures of creativity—review and critique. *Journal of Personality Assessment, 45*, 450–460.

Hoffman, D. H., Greenberg, P., & Fitzner, D. H. (1980). *Lifelong learning and the visual arts: A book of readings.* Weston, VA: National Art Education Association.

Hoffman, Joseph. (1965). Giorgione's three ages of man. *Pantheon, 42*, 238–244.

Holahan, C. K., Sears, R. R., & Cronbach, L. J. (1995). *The gifted group in later maturity.* Stanford, CA: Stanford University Press.

Holden, S. (1999, January 20). When winning isn't the only thing. *The New York Times,* B14.

Hood, W. (Ed.). (1977–78). Artists and old age: A symposium. *Bulletin,* Allen Memorial Art Museum, Oberlin College, *35*, 2–77.

Horn, J. L. (1982). The aging of human abilities. In B. B. Wolman (Ed.), *Handbook of developmental psychology* (pp. 847–870). Englewood Cliffs: Prentice-Hall.

Horn, J. L., & Cattell, R. B. (1996). Refinement and test of the theory of fluid and crystalline intelligence. *Journal of Educational Psychology, 57*, 253–270.

Horner, K. L., Rushton, J, P., & Vernon, P. A. (1986). Relation between aging and research productivity of academic psychologists. *Psychology and Aging, 1*, 319–324.

Howe, M. J. A. (1982). Biographical evidence and the development of outstanding individuals. *American Psychologist, 37*, 1071–1081.

Hubbard, Linda. (1983). Partners in learning. *Modern Maturity, 26*, 87–88.

Hufford, Mary. (1987). *The grand generation: Memory, mastery and aging.* Washington, DC: Smithsonian.

(1980). [Interrelationships between art and science (and technology)—A bibliography.] *Leonardo, 13*, 29–33.

Jamison, K. R. (1994). *Touched with fire: Manic depressive illness and the artistic temperament.* NY: Free Press.

Jankovitz, F. R. (1977). The senior citizen cultural revolution. *Aging, 3,* 12–16.

Janson, H. W. (Ed.). (1959). *Key monuments of the history of art: A visual survey.* Englewood Cliffs, NJ: Prentice-Hall; NY: Harry Abrams.

Jacques, M. (1980). *Images of age.* Cambridge, MA: Abt Books.

Jaques, E. (1965). Death and the life crisis. *International Journal of Psychoanalysis, 46,* 502–514.

Jaquish, G. A., & Ripple, R. E. (1981). Cognitive creative abilities and self-esteem across the adult life-span. *Human Development, 24,* 110–119.

Johnson, M. M. S. (1990). Age differences in decision making: A process methodology for examining strategic information processing. *Journal of Gerontology, 45,* 75–78.

Johnson, A. C., Prieve, E. A., & others (1976?). *Older Americans: An unrealized audience for the arts.* University of Wisconsin: Madison Center for Arts Administration.

John-Steiner, V. (1995). *Notebooks of the mind: Explorations of thinking.* Albuquerque: University of New Mexico Press.

Kastenbaum, R. (1989). Old men created by artists: Time and transcendence in Tennyson and Picasso. *International Journal of Aging and Human Development, 28,* 81–104.

Kastenbaum, R. (1991a). The creative impulse: Why it won't just quit. In R. Kastenbaum (Ed.), Creativity in later life. *Generations: Journal of the American Society on Aging, 15,* 7–12.

Kastenbaum, R. (Ed.). (1991b). Creativity in later life. *Generations: Journal of the American Society on Aging, 15,* 5–61.

Kastenbaum, R. (1991c). Serious play and infinite limits. In R. Kastenbaum (Ed.), Creativity in later life. *Generations: Journal of the American Society on Aging, 15,* 5–6.

Kastenbaum, R. (1992). The creative process: A life span approach. In T. R. Cole, D. D. Van Tassel, & R. Kastenbaum (Eds.), *Handbook of the humanities and aging* (pp. 285–306). NY: Springer.

Kastenbaum, Robert. (1994a). *Defining acts: Aging as drama.* Amityville, NY: Baywood.

Kastenbaum, R. (1994b). Saints, sages, and sons of bitches: Three models for the grand old man. *Journal of Geriatric Psychiatry, 27,* 61–78.

Kauppinen, H. (1991). Aging in art: Beyond stereotypes. *International Journal of Aging and Human Development, 33,* 217–232.

Kauppinen, H., & McKee, P. L. (1988). Old age painting and gerontology. *Journal of Aesthetic Education, 22,* 86–100.

Keever, J. (1990, September 9). A creature of habit [on Michener]. *The Blade,* Toledo Ohio (page unknown).

Kelly, J. R., Steinkemp, M. W., & Kelly, J. R. (1986). Later life leisure: How they play in Peoria. *Gerontologist, 26,* 531–537.

Kemp, M. (1987). Late Leonardo: Problems and implications. In R. Rosand (Ed.), Old-age style. *Art Journal, 46,* 94–102.

Kleemeier, Robert, W. (Ed.). (1961). *Aging and leisure: A research perspective into the meaningful use of time.* NY: Oxford University Press.

Klein, Julie, T. (1990). *Interdisciplinarity: History, theory, and practice.* Detroit: Wayne State University Press.

Koestler, A. (1964). *Act of creativity.* NY: Macmillan.

Kogan, N. (1973). Creativity and cognitive style: A life-span perspective. In P. B. Baltes & K. W. Schaie (Eds.), *Life-span development psychology* (pp. 145–178). NY: Academic Press.

Kogan, N. (1987). Creativity. In G. L. Maddox (Ed.), *The encyclopedia of aging* (pp. 153–155). NY: Springer.

Labouvie-Vief, G. (1980). Adaptive dimensions of adult cognition. In N. Datan & N. Lohmann (Eds.), *Transitions of aging* (pp. 3–26). NY: Academic Press.

Labouvie-Vief, G. (1984). Logic and self-regulation from youth to maturity: A model. In M. L. Commons, F. A. Richards, & C. Armon (Eds.), *Beyond formal operations: Late adolescent and adult cognitive development* (pp. 158–179). NY: Praeger.

Lambert, D. (1984, January–February). Reflections of genius. *Saturday Review*, 22–25.

Lavin, M. A. (1967). *Vasari, Georgio: Lives of the most eminent painters.* NY: The Heritage Press.

Lehman, H. C. (1953). *Age and achievement.* Princeton, NJ: Princeton University Press/American Philosophical Society.

Lehman, H. C. (1956). Reply to Dennis. *Journal of Gerontology, 11,* 333–337.

Lehman (1958). The influence of longevity upon curves showing man's creative production rate at successive age levels. *Journal of Gerontology, 13,* 187–191.

Lehman, H. C. (1962a). Age of starting to contribute versus total creative output. *Journal of Applied Psychology, 30,* 460–480.

Lehman, H. C. (1962b). More about age and achievement. *Gerontologist, 2,* 141–148.

Lehman, H. C. (1966a). The most creative years of engineers and other technologists. *Journal of Genetic Psychology, 108,* 263–270.

Lehman, H. C. (1966b). The psychologists' most creative years. *American Psychology, 21,* 363–369.

Levin, S., & Kahana, R. J. (1967). *Psychodynamic studies on aging: Creativity, reminiscing, and dying.* NY: International Universities Press.

Levy, B., & Langer, E. (1999). Aging. In M. A. Runco & S. R. Spitzker (Eds.), *Encyclopedia of creativity* (Vol. 1, pp. 45–52). San Diego: Academic Press.

Lindauer, M. S. (1968). Pleasant and unpleasant emotions in literature. *Journal of Psychology, 70,* 55–67.

Lindauer, M. S. (1970a). The effect of clues in perceiving the "good figure." *Perceptual and Motor Skills, 30,* 588.

Lindauer, M. S. (1970b). Psychological aspects of form perception in abstract art. *Science de l'Art, 7,* 19–24.

Lindauer, M. S. (1973). Toward a liberalization of experimental aesthetics. *Journal of Aesthetics and Art Criticism, 31,* 459–465.

Lindauer, M. S. (1975). *The psychological study of literature.* Chicago: Nelson-Hall.

Lindauer, M. S. (1978). Psychology as a humanistic science. *Psychocultural Review: Interpretations in the Psychology of Art, Literature and Society, 2,* 139–145.

Lindauer, M. S. (1982). Psychological aesthetics, aesthetic perception, and the aesthetic person. In D. O'Hare (Ed.), *Psychology and the arts* (pp. 29–75). Sussex, England: Harvester.

Lindauer, M. S. (1983). Imagery and the arts. In A. A. Sheikh (Ed.), *Imagery: Current theory, research, and applications* (pp. 468–506). NY: Wiley.

Lindauer, M. S. (1984a). Applying empirical research methods to the psychology of literature. In J. P. Natoli (Ed.), *Psychological perspectives on literature* (pp. 246–266). NY: Anchor.

Lindauer, M. S. (1984b). Experimental aesthetics. In R. J. Corsini (Ed.), *Wiley encyclopedia of psychology* (Vol. 1, pp. 464–465). NY: Wiley.

Lindauer, M. S. (1984c). Physiognomic perception. In R. J. Corsini (Ed.), *Wiley encyclopedia of psychology* (Vol. 1, pp. 34–35). NY: Wiley.

Lindauer, M. S. (1984d). Physiognomy and art: Approaches from above, below, and sideways. *Visual Arts Research, 10,* 52–65.

Lindauer, M. S. (1984e). Psychology and literature. In M. H. Bornstein (Ed.), *Psychology and its Allied Disciplines* (pp. 113–154). Hillside, NJ: Erlbaum.

Lindauer, M. S. (1986a). Perceiving, imaging, and preferring the colors of physiognomic stimuli. *American Journal of Psychology, 99,* 233–255.

Lindauer, M. S. (1986b). Seeing and touching aesthetic objects: II. Descriptions. *Bulletin of the Psychonomic Society, 24,* 125–126.

Lindauer, M. S. (1987a). Perceived and preferred orientations of abstract art. *Empirical Studies of the Arts, 5,* 47–58.

Lindauer, M. S. (1987b). The psychology of literature and the short story: A methodological perspective. In L. Halasz (Ed.), *Literary discourse: Aspects of cognitive and social psychological approaches* (pp. 125–139). Berlin: de Gruyter.

Lindauer, M. S. (1988). Physiognomic meanings in the titles of short stories. In C. Martindale (Ed.), *Psychological approaches to the study of literary narratives* (pp. 74–95). Hamburg: Helmut Buske.

Lindauer, M. S. (1990). The meanings of the physiognomic stimuli *taketa* and *maluma. Bulletin of the Psychonomic Society, 28,* 47–50.

Lindauer, M. S. (1991a). Physiognomy and verbal synesthesia: Affective and sensory descriptions of nouns with drawings and art. *Metaphor and Symbolic Activity, 6,* 183–202.

Lindauer, M. S. (1991b). Physiognomic expression in literary materials: Bridging the humanities and science. Paper presented at the Society for Science and Literature Conference, Montreal, Canada.

Lindauer, M. S. (1992). Creativity in aging artists: Contributions from the humanities to the psychology of old age. *Creativity Research Journal, 5,* 211–231.

Lindauer, M. S. (1993a). The old-age style and its artists. *Empirical Studies of Aesthetics, 11,* 135–146.

Lindauer, M. S. (1993b). The span of creativity among long-lived historical artists. *Creativity Research Journal, 6,* 221–239.

Lindauer, M. S. (1994). Are creative writers mad? An empirical perspective. In B. M. Rieger (Ed.), *Dionysius in literature: Essays on literary madness.* Bowling Green, OH: Bowling Green State University Popular Press.

Lindauer, M. S. (1995). Psychology, art, and a new look at interdisciplinarity: A personal view. Presidential Address, Division 10, 103rd Annual Meeting of the American Psychological Association, August. Also in *Newsletter of Psychology and the Arts, Division 10, American Psychological Association, Fall/Winter,* 12–16.

Lindauer, M. S. (1998a) Artists, art, and arts activities: What do they tell us about aging? In C. Adams-Price (Ed.), *Creativity and successful aging: Theoretical and empirical approaches* (pp. 237–250). NY: Springer.

Lindauer, M. S. (Ed.). (1998b). Interdisciplinarity, the psychology of art, and creativity [Special issue]. *Creativity Research Journal, 11,* 1–77.

Lindauer, M. S. (1999). Old age style. In M. S. Runco & S. Pritzker (Eds.), *Encyclopedia of creativity* (Vol. 2, pp. 311–318). San Diego: Academic Press.

Lindauer, M. S. (n.d.). *Science, art, and the humanities: Bridging the disciplines through the psychology of art.* Unpublished manuscript.

Lindauer, M. S., & Long, D. (1986). The criteria used to judge art: Marketplace and academic comparisons. *Empirical Studies in the Arts, 4,* 163–174.

Lindauer, M. S., Orwoll, L., & Kelley, C. (1997). Aging artists on the creativity of their old age. *Creativity Research Journal, 10,* 133–152.

Lindauer, M. S., & Ribner, S. (1976). *The arts in correctional settings: I: An interpretive survey. II: A summary of a program of research.* Brockport, NY: Unpublished report.

Lindauer, M. S., Stergiou, E., & Penn, D. (1986). Seeing and touching aesthetic objects: I. Judgments. *Bulletin of the Psychonomic Society, 24,* 121–124.

Linksz, A. (1980). *An ophthalmologist looks at art.* San Francisco, CA: Smith-Kettlewell Eye Research Foundation.

Lyell, R. G. (Ed.). (1980). *Middle age, old age: Short stories, plays, and essays on aging.* NY: Harcourt, Brace Jovanovitch.

McCrae, R. R. (1999). Consistency of creativity across the life span. Aging. In M. A. Runco & S. R. Spitzker (Eds.), *Encyclopedia of creativity* (Vol. 1, pp. 361–366). San Diego: Academic Press.

McCrae, R. R., Arenberg, D., & Costa, P. T., Jr. (1987). Declines in divergent thinking with age: Cross-cultural, longitudinal, and cross-segmental analyses. *Psychology and Aging, 2,* 130–137.

McCutcheon, P. (1965). *Resource guide to people, places and programs in arts and aging.* Washington, DC: National Council on the Aging, National Center Arts and Aging.

McCutcheon, P., & Tecot, K. (1985). *Developing older audiences: Guidelines for performing arts groups.* Washington, DC: National Council on the Aging, National Center on Arts and the Aging.

McDonald, K. A. (1997). Doubling of population unlikely by 2100. *The Chronicle of Higher Education, 43*(43), A16.

McKee, P. L., & Kauppinen, (1987). *The art of aging: A celebration of old age in the history of Western art.* NY: Human Sciences Press.

MacKinnon, D. W. (1962). The nature and nurture of creative talent. *American Psychologist, 57,* 484–495.

MacKinnon, D. W. (1964). The creativity of architects. In C. W. Taylor (Ed.), *Widening horizons in creativity* (pp. 359–378). NY: Wiley.

McLeish, J. A. B. (1976). *The Ulyssean adult: Creativity in the middle and later years.* NY: McGraw-Hill/Ryerson.

McLerrran, J., & McKee, P. (1991). *Old age in myth and symbols. A cultural dictionary.* NY: Greenwood.

McManus, I. C. (1975). Life expectation of Italian Renaissance artists. *The Lancet, 61,* 266–267.

McMurray, J. (Ed.), (1989). Creative arts with older people. *Activities, Adaptation and Aging, 14,* 1–138.

Maddox, G. L. (Ed.). (1987). *The encyclopedia of aging.* NY: Springer-Verlag.

Malcahy, K. V., & others. (1980). The government, art, and aesthetic education. *Journal of Aesthetic Education, 14,* 5–54.

Man and his years: An exhibition presented by the Baltimore Museum of Art (1954). Baltimore: Baltimore Museum of Art.

Manheimer, Ronald, J. (1984). *Developing arts and humanities programming with the elderly.* Chicago, IL: References and Adult Services Division, American Library Association.

Manheimer, R. J. (1987–88). The politics and promise of cultural enrichment programs. Late life learning [special issue]. *Generations, 12,* 20–30.

Martindale, C. (1975). *The romantic progression: The psychology of literary history.* NY: Hemisphere.

Martindale, C. (1995). Creativity and connectionism. In S. M. Smith, T. B. Ward, & R. A. Finke (Eds.), *The creative cognition approach* (pp. 249–268). Cambridge, MA: Bradford/MIT.

Martindale, C. (1999). Art and aesthetics. In M. A. Runco & S. R. Spitzker (Eds.), *Encyclopedia of creativity* (Vol. 1, pp. 115–120). San Diego: Academic Press.

Mayer, R. (1999). Problem solving. In M. A. Runco & S. R. Spitzker (Eds.), *Encyclopedia of creativity* (Vol. 2, pp. 437–447). San Diego: Academic Press.

Montagu, A. (1989). *Growing old* (2nd ed.). Granby, MA: Bergin and Garbey.

Moody, Harry R., Jr. (1982a). *Abundance of life: Human development policies for an aging society*. NY: Columbia University.

Moody, H. R. (1982b), *Aging and cultural policy*. Washington, DC: Washington Council on the Aging.

Moody, H. R. (1987). Humanities and the arts. In G. L. Maddox (Ed.), *The encyclopedia of aging* (pp. 338–341). NY: Springer-Verlag.

Mullin, H., Nolton, B., & others (1990). The seasoned eye 3: Winning works of art. *Modern Maturity*, *33*(3–5), June–July, 46–53, August–September, 42–51, and October– November, 48–53.

Mumford, M. D. (1984). Age and outstanding occupational achievement: Lehman revisited. *Journal of Vocational Behavior*, *25*, 225–244.

Mumford, M. D., & Gustatson, S. B. (1988), Creativity syndrome: Integration, application and innovation. *Psychological Bulletin*, *103*, 27–43.

Munnichs, J. M. (1990). The senior scholar: A three-sided mirror for society. *Gerontology and Geriatrics Education*, *11*, 53–66.

Munsterberg, H. (1983). *The crown of life: Artistic creativity in old age*. NY: Harcourt Brace Jovanovich.

Munsterberg, H. (1994). The critic at seventy-five. In R. Berlind (Ed.), Art and old age. *Art Journal*, *53*, 64–65.

Murray, P., & Murray, L. (1965). *Dictionary of art and artists*. NY: Praeger.

Murray, B., Sleek, S., & DeAngeles, T. (1997). The aging revolution. *American Psychological Association Monitor*, *28*, 22–26.

The myth and reality of aging. (1978). Washington, DC: National Council on Aging: Inter-University Consortium for Political and Social Research.

Nelson, H. (1928). The creative years. *American Journal of Psychology*, *40*, 303–311.

Neugarten, Bernice L., et al. (1980). *Personality in middle and late life*. NY: Arno. (Originally published 1964.)

Nystrom, H. (1979). *Creativity and innovation*. Chichester, England: John Wiley & Sons.

O'Quinn, K., & Besemer, S. P. (1999). Creative products. In M. A. Runco & S. R. Spitzker (Eds.), *Encyclopedia of creativity* (Vol. 1, pp. 413–422). San Diego: Academic Press.

O'Reilly, P., & Caro, F. G. (1994). Productive aging: An overview of the literature. *Journal of Aging & Social Policy*, *6*, 39.

Orwoll, L., & Kelley, M. C. (1998). Personal force and symbolic reach in older women artists. In C. Adams-Price (Ed.), *Creativity and aging: Theoretical and empirical perspectives* (pp. 175–193). NY: Springer.

Park, D. C. (1992). Applied cognitive age research. In Fergus I. M. Craik & Timothy A. Salthouse (Eds.), *The handbook of aging and cognition* (pp. 449–493). Hillsdale, NJ: Erlbaum.

Park, D. C., Royal, D., Dudley, W., & Morrell, W. (1988). Forgetting of pictures over a long retention interval in old and young adults. *Psychology and Aging*, *3*, 94–95.

Parnes, Sidney, J. (1999). Programs and courses in creativity. In M. A. Runco & S. R. Spitzker (Eds.), *Encyclopedia of creativity* (Vol. 2, pp. 465–477). San Diego: Academic Press.

Parsons, M. J. (1987). *How we understand art: A cognitive development account of aesthetic experience*. Cambridge, England: Cambridge University Press.

Perlmutter, M. (Ed.). (1990). *Late life potential*. Washington, DC: The Gerontological Society of America.

Petrakis, E., & Hasnon, C. J. (1981). Cognitive style and choice of leisure activities. *Perceptual & Motor Skills, 52,* 839–842.

Pettys, C. (1985). *Dictionary of women artists: An international dictionary of women artists born before 1900.* Boston: O. K. Hall & Co.

Polansky, G. (1973). Age and creativity. In E. Pfeiffer (Ed.), *Successful aging: A conference report* (pp. 109–111). Durham, NC: Duke University Press.

Pollock, G. H. (1984). Introduction: Aging and creativity. In J. E. Gedo & G. H. Pollock (Eds.), *Psychoanalysis: The vital issues. Vol. 1. Psychoanalysis as an intellectual discipline. Emotion and Behavior Monographs,* No. 2 (pp. 257–275). NY: International Press.

Poon, L. W., Rubin, D. C., & Wilson, B. A. (Eds.). (1989). *Everyday cognition in adulthood and later life.* Cambridge: Cambridge University Press.

Pribic, E. (Ed.). (1990). *Nobel laureates in literature: A biographical dictionary.* NY: Garland.

Price, D. J. de Solla. (1963). *Little science, big science.* NY: Columbia University Press.

Pritikin, R. (1990). Marcel Duchamp, the artist, and the social expectations of aging. *The Gerontologist, 30,* 636–639.

Pruyser, P. W. (1987). Creativity in aging persons. *Bulletin of the Menninger Clinic, 51,* 425–435.

Pufal-Struzik, I. (1992). Differences in personality and self-knowledge of creative persons of difference ages: A comparative analysis. Geragogics: European research in gerontological education and educational gerontology [Special issue]. *Gerontology and Geriatrics Education, 13,* 71–90.

Qualls, Ss. H., & Abeles, N. (Ed.). (2000). *Psychology and the aging revolution: How we adapt to longer life.* Washington, DC: American Psychological Association.

Rabasca, L. (1999). Happiness may increase with age. *American Psychological Association Monitor, 30,* 11.

Raina, M. K. (1999). Cross-cultural differences. In M. A. Runco & S. R. Spitzker (Eds.), *Encyclopedia of creativity* (Vol. 1, pp. 453–464). San Diego: Academic Press.

Rankin, S. H. (1989). The emergence of creativity in later life. In R. T. Mercer, E. G. Nichols, & G. C. Doyle (Eds.), *Transitions in a woman's life: Major life events in developmental context* (pp. 167–178).

Raskin, E. (1936). Comparisons of scientific and literary ability: A biographical study of eminent scientists and men of letters of the nineteenth century. *Journal of Abnormal and Social Psychology, 31,* 20–35.

Ravin, J. G. (1985). Monet's cataracts. *Journal of the American Medical Association, 253,* 394–399.

Ravin, J. G., & Kenyon, C. A. (1998). Artistic vision in old age: Claude Monet and Edgar Degas. In C. Adams-Price (Ed.), *Creativity and aging: Theoretical and empirical perspectives* (pp. 251–267). NY: Springer.

Raymond, C. (1989, June 21). Creative works by neurologically impaired artists provide scientists with a window on the brain. *The Chronicle of Higher Education, 35,* A4–6.

Reiss, S. M. (1999). Women and creativity. In M. A. Runco & S. R. Spitzker (Eds.), *Encyclopedia of creativity* (Vol. 2, pp. 699–708). San Diego: Academic Press.

Richards, R. (1999a). Everyday creativity. In M. A. Runco & S. R. Spitzker (Eds.), *Encyclopedia of creativity* (Vol. 1, pp. 683–687). San Diego: Academic Press.

Richards, R. (1999b) Four Ps of creativity. In M. A. Runco & S. R. Spitzker (Eds.), *Encyclopedia of creativity* (Vol. 1, pp. 733–742). San Diego: Academic Press.

Richardson, J. (1997). The great forgotten modernist. *The New York Review of Books, 64,* 31–32.

Ripple, Richard, E. (1999). Teaching creativity. In M. A. Runco & S. R. Spitzker (Eds.), *Encyclopedia of creativity* (Vol. 2, pp. 629–638). San Diego: Academic Press.

Rodeheaver, D. (1987). Problem solving. In G. L. Maddox (Ed.), *The encyclopedia of aging* (pp. 537–539). NY: Springer-Verlag.

Rodeheaver, D., Emmons, C., & Powers, K. (1998). Context and identify in women's late life creativity. In C. Adams-Price (Ed.), *Creativity and aging: Theoretical and empirical perspectives* (pp. 195–234). NY: Springer.

Roe, A. (1972). Maintenance of creative output through the years. In C. W. Taylor (Ed.), *Climate for creativity* (pp. 167–191). NY: Pergamon.

Roediger, H. L. III. (1990). Implicit memory: Retention without remembering. *American Psychologist, 45*, 1043–1056.

Romaniuk, J. G., & Romaniuk, M. (1981). Creativity across the life span: A measurement perspective. *Human Development, 24*, 366–381.

Root-Bernstein, R. (1999). Productivity and age. In M. A. Runco & S. R. Spitzker (Eds.), *Encyclopedia of creativity* (Vol. 2, pp. 457–463). San Diego: Academic Press.

Root-Bernstein, R. S., Bernstein, M., & Garnier, H. (1993). Identification of scientists making long-term high-impact contributions, with notes on their method of working. *Creativity Research Journal, 6*, 329–343.

Rosand, D. (Ed.). (1987). Old-age style. *Art Journal, 46*, 91–133.

Rosand, D. (1990, October 28). The challenge of Titian's "Senile Sublime," *The New York Times, 140 H41*, 44–47.

Rose, L. H., & Lin, H. T. (1984). A meta-analysis of long-term creativity training programs. *Journal of Creative Behavior, 18*, 11–22.

Rosenberg, J. (1980) *Rembrandt: Life and work* (2nd ed.). Ithaca, NY: Cornell University/Phaidon. (Originally published 1964.)

Rosenblum, R. (1977–78). Mondrian's late style. *Allen Memorial Art Museum Bulletin Oberlin College, 35*, 68–77.

Rosenthal, G. (1954). Introduction. In *Man and his years: Catalogue of an exhibition by the Baltimore Museum of Art* (pp. 4–12). Baltimore, MD: Baltimore Museum of Art.

Rosenthal, G. (Ed.). (1968). *From El Greco to Pollock: Early and late works by European and American artists.* Greenwich, CN: The Baltimore Museum of Art and the New York Graphic Society.

Roth, Henry. (1988). Contemporary authors: New revision series. 63. Detroit, MI.

Rothenberg, A. (1996). The Janusian process in scientific creativity. *Creativity Research Journal, 9*, 207–231.

Rothenberg, A., & Greenberg, B. (1974). *The index of scientific writings on creativity.* Archon Books/The Shoe String Press.

Rowe, J. W., & Kahn, R. L. (1987). Human aging: Usual and successful. *Science, 237*, 143–149.

Rowe, J. W., & Kahn, R. L. (1997). Successful aging. *The Gerontologist, 37*, 433–440.

Rowe, J. W., & Kahn, R. L. (1998), *Successful Aging.* NY: Pantheon Books.

Runco, M. A. (1995). Insight for creativity, expression for impact, *Creativity Research Journal, 8*, 377–390.

Runco, M. A. (1999). Tests of creativity. In M. A. Runco & S. R. Spitzker (Eds.), *Encyclopedia of creativity* (Vol. 2, pp. 755–760). San Diego: Academic Press.

Runco, M. A., & Albert, R. S. (Eds.). (1990). *Theories of creativity.* Newbury Park: Sage.

Runco, M. A., & Dow, G. (1999). Problem finding. In M. A. Runco & S. R. Spitzker (Eds.), *Encyclopedia of creativity* (Vol. 2, pp. 433–435). San Diego: Academic Press.

Runco, M. A., & Spitzker, S. R. (Eds.). (1999). *Encyclopedia of creativity* (2 vols). San Diego: Academic Press.

Ruth, J. E., & Birren, J. E. (1985). Creativity in adulthood and old age: Relations to intelligence, sex, and mode of testing. *Behavioral Development, 8*, 99–109.

Ruth, J. E., & Westerlund, C. (1979). Age and creativity. *Year Book Gerontology, 22*, 47–58
(Helsinki).

Ryff, C. D. (1982). Successful aging: A developmental approach. *The Gerontologist, 22*,
209–214.

Salthouse, T. A. (1987). Perception. In G. L. Maddox (Ed.), *The encyclopedia of aging*
(pp. 517–518). NY: Springer-Verlag.

Salthouse, T. A. (1989). Age-related changes in basic cognitive processes. In M. Storandt
& G. R. Vander Bos (Eds.), *The adult years: Continuity and change* (pp. 9–40).
Washington, DC: American Psychological Association.

Salthouse, T. A. (1990). Cognitive competence and expertise in aging. In J. E. Birren and
K. W. Schaie (Eds.), *Handbook of the psychology of aging* (3rd ed., pp. 311–319).
San Diego: Academic Press.

Sampson, A., & Sampson, S. (1985). *The Oxford book of ages*. NY: Oxford University
Press.

Sasser-Cohen, J. R. (1993). Qualitative changes in creativity in the second half of life:
A life-span developmental perspective. *Journal of Creative Behavior, 27*,
18–27.

Sayre, E. A. (1956). An old man's writing: A study of Goya's albums. *Bulletin of the
Museum of Fine Arts, Boston, 56*, 116–136.

Schaie, K. W., (1990). Intellectual development in adulthood. In J. E. Birren &
K. W. Schaie (Eds.), *Handbook of the psychology of aging* (3rd ed., pp. 291–309).
San Diego: Academic Press.

Schaie, K. W., & Stroller, C. A. (1968). A cross-sequential study of age changes in cognitive behavior. *Psychological Bulletin, 70*, 671–680.

Schaie, K. W., & Willis, S. L. (1986). Reintegration or despair. In K. W. Schaie &
S. L. Willis (Eds.), *Adult development and aging* (2nd ed., pp. 449–486). Boston:
Little, Brown.

Schapiro, M. (1953). Style. In A. L. Krober (Ed.), *Anthropology today* (pp. 311–320).
Chicago: University of Chicago.

Schneidman, E. (1989). The Indian summer of life: A preliminary study of septuagenarians. *American Psychologist, 44*, 684–694.

Scotch, S. (1989). Foreword. In J. McMurray (Ed.), Creative arts with older people.
Activities, Adaptation and Aging, 14, 1–138.

Schulz, R., & Heckhausen, J. (1996). A life span model of successful aging. *American
Psychologist, 51*, 702–714.

Sears, R. R., & Feldman, S. S. (Eds.). (1973). *The seven ages of man*. Los Altos, CA:
Kaufman.

Sears, R. (1977). Sources of life's satisfactions of Terman's gifted men. *American
Psychologist, 32*, 719–728.

Seligman, M. E. P., & Csikszentmihalyi, M. (Eds.). (2000). Happiness, excellence, and
optimal human functioning. *American Psychologist, 55*, 5–183.

Setlow, C. E. (1984). *Older Americans and the arts*. Washington, DC: National Council on
Aging.

Shenk, D., & Achenbaum, W. A. (Eds.). (1994). *Changing perceptions of aging and the
aged*. NY: Springer.

Shimamura, A. P., Berry, J. M., Mangels, J. A., & Rusting, C. L. et al. (1995). Memory
and cognitive abilities in university professors: Evidence for successful aging.
Psychological Science, 6, 271–277.

Silbergeld, J. (1987). Chinese concepts of old age and their role in Chinese painting, poetry, theory, and criticism. *Art Journal, 46,* 103–114.

Simonton, D. K. (1976a). Philosophical eminence, beliefs, and Zeitgeist: An individual-generational analysis. *Journal of Personality and Social Psychology, 34,* 630–640.

Simonton, D. K. (1976b). Ideological diversity and creativity: A reexamination of a hypothesis. *Social Behavior and Personality, 4,* 203–207.

Simonton, D. K. (1977). Creative productivity, age and stress: A biographical time-series analysis of 10 classical composers. *Journal of Personality and Social Psychology, 35,* 794–804.

Simonton, D. K. (1981). Creativity in Western civilization. Intrinsic and extrinsic causes. *American Anthropologist, 83,* 628–630.

Simonton, D. K. (1984a). Artistic creativity and interpersonal relations across and within generations. *Journal of Personality and Social Psychology, 46,* 1273–1286.

Simonton, D. K. (1984b). Creative productivity: A mathematical model. *Developmental Review, 4,* 77–121.

Simonton, D. K. (1984c). *Genius, creativity, and leadership: Historiometric inquiries.* Cambridge, MA: Harvard University Press.

Simonton, D. K. (1986). Developmental antecedents of achieved eminence. In *Annals of child development* (Vol. 4). Greenwich, CT: JAI Press.

Simonton, D. K. (1988a). Age and outstanding achievement: What do we know after a century of research? *Psychological Bulletin, 104,* 251–267.

Simonton, D. K. (1988b). Creativity, leadership, and chance. In R. J. Sternberg (Ed.), *The nature of creativity* (pp. 386–426). Cambridge, MA and NY: Cambridge University Press.

Simonton, D. K. (1989). The swan-song phenomenon: Last-work effects for 172 classical composers. *Psychology and Aging, 4,* 42–47.

Simonton, D. K. (1990a). Creativity and wisdom in aging. In J. E. Birren & K. W. Schaie (Eds.), *Handbook of the psychology of aging* (3rd ed., pp. 320–329). NY: Academic Press.

Simonton, D. K. (1990b). Does creativity decline in the later years? In M. Perlmutter (Ed.), *Late life potential* (pp. 83–112). Washington, DC: The Gerontological Society of America.

Simonton, D. K. (1990c). *Scientific genius: A psychology of science.* NY: Cambridge University Press.

Simonton, D. K. (1991a) Creative productivity through the adult years. In R. Kastenbaum (Ed.), Creativity in later life. *Generations: Journal of the American Society on Aging, 15,* 13–16.

Simonton, D. K. (1991b). Creativity in the later years: Optimistic prospects for achievement. *The Gerontologist, 30,* 626–634.

Simonton, D. K. (1994). *Greatness: Who makes history and why?* NY: Guilford.

Simonton, D. K. (1997). Creative productivity: A prediction and explanatory model of career trajectories and landmarks. *Psychological Review, 104,* 66–89.

Simonton, D. K. (1998). Masterpieces in music and literature: Historiometric inquires: Arnheim Award Address to Division 10 of the American Psychological Association. *Creativity Research Journal, 11,* 103–110.

Simonton, D. K. (1999a). Eminence. In M. A. Runco & S. R. Spitzker (Eds.), *Encyclopedia of creativity* (Vol. 1, pp. 647–657). San Diego: Academic Press.

Simonton, D. K. (1999b). Historiometry. In M. A. Runco & S. R. Spitzker (Eds.), *Encyclopedia of creativity* (Vol. 2, pp. 815–822). San Diego: Academic Press.

Sinnott, J. D. (1998). Creativity and postformal thought: Why the last stage is the creative stage. In C. Adams-Price (Ed.), *Creativity and aging: Theoretical and empirical perspectives* (pp. 43–72). NY: Springer.

Smith, A. D., Park, D. C., Cherry, K., & Berkovsky, K. (1990). Age differences in memory for concrete and abstract pictures. *Journal of Gerontology, 45,* 205–209.

Smith, G. J. W. (1989). Creativity as a key factor. *Psychological Research Bulletin, 29,* 1–24.

Smith, G. J., & Kragh, U. (1975). Creativity in males and old age. *Psychological Research Bulletin, 15,* 1–17.

Smith, S. M., Ward, T. B., & Finke, R. A. (Eds.). (1995). *The creative cognition approach.* Cambridge, MA: Bradford/MIT.

Snow, C. P. (1959). *The two cultures and the scientific revolution.* NY: Cambridge.

Solomon, R., Powell, K., & Gardner, H. (1999). Multiple intelligences. In M. A. Runco & S. R. Spitzker (Eds.), *Encyclopedia of creativity* (Vol. 1, pp. 273–283). San Diego: Academic Press.

Soussloff, C. M. (1987). Old age and old-age style in the "Lives" [a 16th–17th century book] of artists: Gianlorenzo Bernini. *Art Journal, 46,* 122–126.

Spaniol, S. E. (1992). Is there a "late style of art? Line use in human figure drawings by elderly people. *Art Therapy, 9,* 93–95.

Spencer, J. R. (1962). Introduction. In Youthful works by great artists: A symposium. *Allen Memorial Art Museum Bulletin Oberlin College, 20,* 90–91.

Stephens, S. (1998). Accomplished in all departments of art. Review of James F. O'Gorman, Hammatt Billings of Boston, 1818–1874. *The New York Times Book Review, 38,* 25–26.

Steinberg, David. (2001, January 14). For writers, life just begins at 50. *San Francisco Chronicle,* 18.

Sternberg, R. J. (1985). Implicit theories of intelligence, creativity and wisdom. *Journal of Personality and Social Psychology, 49,* 607–627.

Sternberg, R. J. (Ed.). (1988). *The nature of creativity: Contemporary psychological perspectives.* Cambridge, MA and NY: Cambridge University Press.

Sternberg, R. J., & Dees, N. K. (Eds.). (2001). Creativity for the new millennium [Special issue]. *American Psychologist, 56,* 332–362.

Sternberg, R. J., Wagner, R. K., Williams, W. M., & Horvath, J. A. (1995). Testing common sense. *American Psychologist, 50,* 912–926.

Stone, D. L. (1991). Self-perceptions of older adults as artists. *Activities, Adaptation & Aging, 16,* 31–37.

Subotnik, R. F., & Arnold, K. D. (Eds.). (1994). *Beyond Terman: Contemporary longitudinal studies of giftedness and talent.* Stanford, CT: Ablex Publishing Co.

Subotnik, R. F., & Arnold, K. D. (1999). Longitudinal studies. In M. A. Runco & S. R. Spitzker (Eds.), *Encyclopedia of creativity* (Vol. 2, pp. 163–165). San Diego: Academic Press.

Sunderland, J. T. (1978). Art's aglow with rekindled interest. *Perspectives on Aging, 7,* 27–28.

Sunderland, J. T. (1982). The arts and the aging advocacy movement in the United States: A historical perspective. *Educational Gerontology, 8,* 195–205.

Sunderland, J. T. (1976). *Older Americans and the arts: A human equation.* Washington, DC: John F. Kennedy Center for the Performing Arts. National Council on the Aging.

Sunderland, J. T., Traylor, N. J., & Peter Smith Associates (Eds.). (1976–77). *Arts and the aging. An agenda for action: A national conference.* Minneapolis, MN: National Council on the Aging/ National Center on the Arts and Aging.

Susan, M. (1984). Realities of aging: Starting points for imaginative work with the elderly. *Journal of Gerontological Social Work, 7,* 201–213.

Syclox, K. (1983). Our elderly: Creatives often overlooked. *Activities, Adaptation, Aging, 3*, 27–28.

Talley, D. R. (1990). Aging: The process, the perception. *James Community College* [Newsletter], Jamestown, NY.

Taunton, M. (1982). Aesthetic responses of young children to the visual arts: A review of the literature. *Journal of Aesthetic Education, 16*, 93–109.

Taylor, I. A. (1959). The nature of the creative process. In P. Smith (Ed.), *Creativity: An examination of the creative process* (pp. 51–82). NY: Hastings House.

Taylor, I. A. (1973). Patterns of creativity and aging. In E. Pfeiffer (Ed.), *Successful aging: A conference report* (pp. 113–117). Durham, NC: Duke University Press.

Taylor, C. W., & Sacks, D. (1981). Facilitating lifetime creative processes—a think piece. *Gifted Child Quarterly, 25*, 116–118.

Terman, L. M., & Oden. M. H. (1959). *The gifted group at mid-life: Thirty-five years' follow-up of the superior child.* Stanford, CA: Stanford University Press.

Terman, L. M., & Oden, M. H. (1925–59). *Genetic studies of geniuses* (Vols. 1–4). Stanford, CA: Stanford University Press.

Tinsley, H. E. A., Colbs, S. L., Teaff, J. D., & Kaufman, N. (1987). The relationship of age, gender, health, and economic benefits older persons report from participation in leisure activities. *Leisure Sciences, 9*, 53–65.

Tietze, H. (1944–46). Earliest and latest works of great artists. *Gazette des Beaux-Arts, 46*, 273–284.

Tomas, V. (1964). *Creativity in the arts.* Englewood Cliffs, NJ: Prentice-Hall.

Torrance, E. P. (1977). Creativity and the older adult. *Creative Child Adult Quarterly, 2*, 136–144.

(1981–84). *Training of service providers in establishing arts and humanities programs for old adults.* St. Louis, MO: Cemrel.

Trevor-Roper, P. (1970). *The world through blunted sight: An inquiry into the influence of defective vision on art and character.* Indianapolis: The Bobbs-Merril Co.

Udell, C. J. (1979). *Humanities program for the elderly: Final report: field survey and literature search.* Radnor, PA: Chilton Research Service; Washington, DC: B'nai B'rith International.

United States Congress, House Select Committee on Aging. Subcommittee on Human Services. (1980, February 7. *The arts and the old American: Hearing before the Subcommittee on Human Services of the Select Committee on Aging*, House of Representatives, Ninety-sixth Congress, Second session).

Useem, M. (1976). Government patronage of science and art in America. *Behavioral Scientist, 19*, 755–804.

Uttal, D. H., & Perlmutter, M. (1989). Toward a broader conceptualization of development: The role of gains and losses across the life span. *Developmental Psychology, 9*, 1–32.

Walberg, H. J., & Arian, G. (1999). Distribution of creativity. In M. A. Runco & S. R. Spitzker (Eds.), *Encyclopedia of creativity* (Vol. 1, pp. 573–576). San Diego: Academic Press.

Walker, J. (Ed.). (1996). *Halliwell's film guide.* NY: HarperPerennial/HarperCollins.

Wass, H., & Olejnik, S. F. (1983). An analysis and evaluation of research in cognition and learning among older adults. *Educational Gerontology, 9*, 323–333.

Wetle, T. (1991). Successful aging: New hope for optimizing mental and physical well-being. *Journal of Geriatric Psychiatry, 24*, 3–12.

White, R. (1977–78). Rubens and old age. *Allen Memorial Art Museum Bulletin Oberlin College, 35*, 40–56.

White, R. R. (1931). Versatility of genius. *Journal of Abnormal and Social Psychology, 2,* 460–489.

(1990) *Who's Who in American art.* NY: R. R. Bowker.

Williams, S. A., Denney, N. W., & Schadler, M. (1983). Elderly adults' perception. *International Journal of Aging and Human Development, 16,* 147–158.

Winkler, M. G. (1992). Walking to the stars. In T. R. Cole, D. D. Van Tassel, & R. Kastenbaum (Eds.), *Handbook of the humanities and aging* (pp. 258–284). NY: Springer.

Winner, E. (1982). *Invented worlds: The psychology of the arts.* Cambridge: Harvard University Press.

Wong, P. T. (1989). Personal meaning and successful aging. Psychology of aging and gerontology [Special issue]. *Canadian Psychology, 30,* 516–525.

Wood, C. (1986). *100 Masterpieces in color.* England: Galley Press/Hamlyn Pub. (Originally published 1972.)

Woodward, K. (1994). Simone de Beavoir: Prospects for the future of older women. In D. Shenk & W. A. Achenbaum (Eds.), *Changing perceptions of aging and the aged* (pp. 31–38). NY: Springer.

Woodward, K., & Schwartz, M. W. (Eds.). (1986). *Meaning and desire: Aging, literature, psychoanalysis.* Bloomington, IN: Indiana University Press.

Wyatt-Brown, A. M. (1988). Late style in the novels of Barbara Pym and Penelope Mortimer. *Gerontologist, 28,* 835–839.

Wyatt-Brown, A. M. (1990). The coming of age of literary gerontology. *Journal of Aging Studies, 4,* 299–315.

Wyatt-Brown, A. M. (1992). Literary gerontology comes of age. In T. R. Cole, D. D. Van Tassel, & R. Kastenbaum (Eds.), *Handbook of the humanities and aging* (pp. 258–284). NY: Springer.

Wyatt-Brown. A. M. (1993). Introduction: Aging, gender, and creativity. In A. M. Wyatt-Brown & J. Rossen (Eds.), *Aging and gender in literature: Studies in creativity* (pp. 1–15). Charlottesville, VA: University Press of Virginia.

Wyatt-Brown, A. M., & Rossen, J. (Eds.). (1993). *Aging and gender in literature: Studies in creativity.* Charlottesville, VA: University Press of Virginia.

Yanke, R. E. (1994). Representations of aging in contemporary literary works. In D. Shenk & W. A. Achenbaum (Eds.), *Changing perceptions of aging and the aged* (pp. 31–52). NY: Springer.

Yanke, R. E., & Eastman, R. M. (1995). *Literature and gerontology: A research guide.* Westport, CN: Greenwood Press.

Young, M. C. (1997). *The Guinness Book of World Records.* Stanford, CT: Guinness Publishing.

(1962). Youthful works by great artists: A symposium. *Allen Memorial Art Museum Bulletin Oberlin College, 35,* 40–56.

Zlatin, H. P., & Nucho, A. O. (1983). The final picture in art. *Journal of Geriatric Psychiatry, 16,* 113–147.

Index